6/14/78   12597

D1480569

PEST CONTROL STRATEGIES

PEST CONTROL STRATEGIES

Edited by

EDWARD H. SMITH

Department of Entomology
Cornell University
Ithaca, New York

DAVID PIMENTEL

Department of Entomology and
Section of Ecology and Systematics
Cornell University
Ithaca, New York

ACADEMIC PRESS   New York   San Francisco   London   1978

A Subsidiary of Harcourt Brace Jovanovich, Publishers

SB 950
.P47

ACADEMIC PRESS, INC.
111 Fifth Avenue, New York, New York 10003

United Kingdom Edition published by
ACADEMIC PRESS, INC. (LONDON) LTD.
24/28 Oval Road, London NW1   7DX

Library of Congress Cataloging in Publication Data

Main entry under title:

Pest control strategies.

"Papers presented at a symposium held at Cornell
University in June 1977."
1.   Pest control—Congresses.   2.   Insect control—
Congresses.   I.   Smith, Edward H.   II.   Pimentel,
David, Date
SB950.A2P46   632'.6   78-109
ISBN 0-12-650450-4

CONTENTS

II   COMPLEXITY OF PEST MANAGEMENT

III   CASE STUDIES OF PEST MANAGEMENT

Potentials for Research and Implementation of Integrated Pest Management on Deciduous Tree-Fruits

Brian A. Croft

Potentialities for Pest Management in Potatoes

H. David Thurston

Insect Control in Corn—Practices and Prospects

William H. Luckmann

IV OBSTACLES AND INCENTIVES

CONTRIBUTORS

Numbers in parentheses indicate the pages on which authors' contributions begin.

EDWARD J. ARMBRUST (85), Illinois Natural History Survey and Illinois Agricultural Experiment Station, Urbana, Illinois 61801

BRIAN A. CROFT (101), Pesticide Research Center and Department of Entomology, Michigan State University, East Lansing, Michigan 48824

M. RUPERT CUTLER (9), Assistant Secretary of Agriculture for Conservation, Research, and Education, USDA, Washington, D.C. 20250

BOYSIE E. DAY (203), Department of Plant Pathology, University of California, Berkeley, California 94720

T. DRLIK (215), Urban Biological Control Project, Division of Biological Control, University of California, Berkeley, California 94720

THOMAS P. GRUMBLY (261), Special Assistant to the Commissioner, Public Health Service, Food and Drug Administration, Rockville, Maryland 20852

ANDREW P. GUTIERREZ (237), Division of Biological Control, University of California, Berkeley, California 94720

DEAN L. HAYNES (181), Department of Entomology, Michigan State University, East Lansing, Michigan 48824

N. HEIDLER (215), Urban Biological Control Project, Division of Biological Control, University of California, Berkeley, California 94720

CARL B. HUFFAKER (237), Division of Biological Control, University of California, Berkeley, California 94720

L. LAUB (215), Urban Biological Control Project, Division of Biological Control, University of California, Berkeley, California 94720

WILLIAM H. LUCKMANN (137), Illinois Natural History Survey and Illinois Agricultural Experiment Station, Urbana, Illinois 61801

M. MINTER (215), Urban Biological Control Project, Division of Biological Control, University of California, Berkeley, California 94720

WARREN R. MUIR (3), Council on Environmental Quality, Washington, D.C. 20006

L. DALE NEWSOM (157), Center for Agricultural Sciences and Rural Development, Louisiana State University, Baton Rouge, Louisiana 70803

RICHARD D. O'BRIEN (23), Division of Biological Sciences, Cornell University, Ithaca, New York 14853

HELGA OLKOWSKI (215), Urban Biological Control Project, Division of Biological Control, University of California, Berkeley, California 94720

WILLIAM OLKOWSKI (215), Urban Biological Control Project, Division of Biological Control, University of California, Berkeley, California 94720

L. ORTHEL (215), Urban Biological Control Project, Division of Biological Control, University of California, Berkeley, California 94720

DAVID PIMENTEL (55), Department of Entomology and Section of Ecology and Systematics, Cornell University, Ithaca, New York 14853

CHRISTINE A. SHOEMAKER (237), Department of Environmental Engineering, Cornell University, Ithaca, New York 14853

EDWARD H. SMITH (309), Department of Entomology, Cornell University, Ithaca, New York 14853

RAY F. SMITH (41), Department of Entomological Sciences, University of California, Berkeley, California 94720

ROGER W. STROHBEHN (73), Natural Resource Economics Division, Economic Research Service, USDA, Washington, D.C. 20250

H. DAVID THURSTON (117), Department of Plant Pathology, Cornell University, Ithaca, New York 14853

R. L. TUMMALA (181), Department of Entomology, Michigan State University, East Lansing, Michigan 48824

RICHARD R. WHETSTONE (271), San Ramon Business Centers, Shell Chemical Company, San Ramon, California 94583

WAYNE R. Z. WILLEY (285), Environmental Defense Fund, Berkeley, California 94704

R. ZUPARKO (215), Urban Biological Control Project, Division of Biological Control, University of California, Berkeley, California 94720

PREFACE

Pest Control Strategies is a collection of papers presented at a symposium held at Cornell University in June 1977. The symposium and resulting book are unique in having assembled some of the nation's leaders in pest control to discuss various strategies for controlling pests. The value of this volume rests with the outstanding contributors and diversity of views on problems of pest control.

The publication of *Pest Control Strategies* comes at an appropriate time when changes in pest control are occurring in the nation. The public has demanded, and rightly so, to see the pesticide use "balance sheet"—to know the risks versus the benefits relative to public health, economics, and the environment. Agriculture is moving to replace routine spraying with treat-when-necessary programs that are based on monitoring of pest and parasite/predator populations. At the same time some research is now being focused on integrating pest control in total agricultural systems management. With this approach, maximum benefits with minimal risks will be possible for agriculture and society.

These presentations are valuable from the standpoint of diversity of approach to pest control and the relevance of the papers to the current situation in the United States; the book as a whole has much to offer students, teachers, and researchers.

ACKNOWLEDGMENTS

We want to acknowledge the stimulating remarks by W. Keith Kennedy (Dean, College of Agriculture and Life Sciences, Cornell University) and Waldemar Klassen (Staff Scientist, Pest Management, Agricultural Research Service, United States Department of Agriculture). Their comments provided a perspective for the symposium and set the stage for stimulating papers presented on a range of topics including the complexity of pest control, specific case studies for pest control in crops, ecosystem management, and obstacles and incentives in integrated pest management.

A book such as this is the product of the efforts and cooperation of a great many people. First, we thank all the contributors, not only for their papers, but also for their prompt help in dealing with minor editorial-production problems. Our sincere thanks are expressed to Ms. Nancy Goodman, Symposium Coordinator, Ms. Judy Hough, and Mr. David Gallahan for their editorial assistance. We also thank Ms. Debra Alesbury, Ms. Beverly DiTaranti, Ms. Alex Laszlo, and Ms. Colleen Martin for their part in organizing the symposium and preparing the manuscripts. We are grateful to Professor Neil Orloff (Program on Science, Technology, and Society) and Professor Raymond Loehr (Environmental Studies Program) of Cornell University, Ms. Maureen Hinkle of the Environmental Defense Fund, and Dr. Michael Jacobsen of the Center for Science in the Public Interest for their role in organizing the symposium. Last and most important, we thank Dr. Warren Muir and Dr. Dale Bottrell of the Council on Environmental Quality and Dr. Edwin L. Johnson and Dr. Kenneth J. Hood of the Environmental Protection Agency for their assistance in the development of the symposium and for their agencies' financial support.

Part I
Introduction

PEST CONTROL—A PERSPECTIVE

Warren R. Muir

Council on Environmental Quality
Washington, D.C.

Synthetic organic pesticides introduced following World War II have brought inestimable benefits to humanity in terms of human lives saved, diminished suffering, and economic gain (1,2). The widescale employment of DDT temporarily resulted in the complete riddance of serious public health pests such as malaria mosquitos from entire countries (3). In the first flush of enthusiasm it even seemed that the broad-spectrum post-World War II pesticides could eliminate pest problems as far ranging as the housefly in the cities and the boll weevil in cotton (4).

These materials were cheap, effective in small quantities, persistent, broadly toxic against whole complexes of pests, and required little labor for their deployment. Thus, it is no wonder that modern pesticides displaced the biological, environmental, and cultural controls which, integrated together and sometimes in combination with the earlier inorganic, botanical, and raw-organic (e.g., creosote) chemical pesticides, formed the prewar pest control arsenal. The new organic synthetic pesticides could be used alone with resulting high levels of pest mortality. Therefore, they eliminated the need for many of the earlier pest control schemes that employed a combination of suppressive measures such as plowing, crop rotations, and selective burning, and were less labor intensive than these integrated schemes.

The miracle postwar pesticides had a spectacular impact on agriculture. As a result of a high return for a low cost investment, the new pesticides were used in large tonnages and rapidly became the primary tools of agricultural pest control on farms of the developed countries. The new pesticides were used to provide season-long pest protection to the crops and to bolster returns from fertilizers and other inputs. They also had a major impact in the

3

underdeveloped countries where the "Green Revolution" continues to occur through the introduction of new high-yielding varieties of wheat, rice, maize, and other food grains (5,6).

The success of pesticides has led to widespread reliance on them, and chemical control has evolved into a highly significant economic activity. In 1971 $3.4 billion worth (retail) of chemical pesticides was applied on a worldwide basis for agricultural (including forestry), industrial, and household use (7). About half was used in the United States where pesticide consumption has upsurged notably in the past 25-30 years. Production of synthetic organic pesticides in this country increased from an estimated 200,000 pounds in 1950 to an estimated 1.4 billion pounds in 1975 (8). During the past decade, the most dramatic pesticide increase has been in the use of herbicides to replace hand labor and machine cultivation in controlling weeds in agricultural crops, in the rights-of-way of highways, utility lines, and railroads, and in cities (9). Agriculture particularly has become dependent on herbicides. Most farming enterprises now rely on herbicides to do the job previously performed by "hoe hands," who have left the farms for city life. Ten years after the first organic herbicide, 2,4-D, had been introduced for large-scale use in 1947, 92% of the farmers in advanced farming areas depended regularly upon this and other herbicides for weed control. It has been said that one worker in a chemical factory producing 2,4-D is equivalent to the work efficiency of 100 hoe hands in a corn field (10).

The unqualified success of chemical pesticides, ironically, created a dilemma. Many of our requisites in life coevolved with pesticide technology to such an extent that we have become, in large part, dependent on this technology. Yet, there are warnings that make us question how secure this technology is for the future.

Increased pest resistance is limiting the effectiveness of many pesticides. Worldwide, there presently are at least 305 species of insects, mites, and ticks that possess strains resistant to one or more chemical pesticides (11). Resistance in certain rodents and plant pathogens is currently limiting the effectiveness of pesticides against these pests (12). We have not yet had to face the development of weed strains resistant to pesticides, at least at levels found in insect pests, and many weed scientists do not view future genetic resistance to herbicides as a serious threat. Nevertheless, weed control literature is full of cases in which control is said to be "difficult" because of decreased effectiveness of the herbicides and many studies have shown the existence of weed strains (ecotypes) relatively resistant to various herbicides (12,13). A recent pest control study

committee of the National Academy of Sciences concluded
"... we believe that the potential for resistance to
herbicides by weeds has been underestimated and recommend
that weed scientists maintain a careful watch for resistance
development" (12).

Despite efforts taken by the Environmental Protection
Agency in the past five years to prohibit or restrict the use
of pesticides that pose intolerable risks, the effects of
pesticides on human health and the environment remain a
continual concern for us all. A study recently completed by
the Council on Environmental Quality suggests a general trend
toward reduced aquatic contamination by chlorinated hydrocarbon insecticides in several major rivers in the southern
United States, even though low levels of these materials
still persist. Difficult to assess are the long-term effects
of the less persistent replacements such as organophosphorus
insecticides.

Finally, the question arises as to how much the growing
petroleum shortage will affect the future of American pest
control whose petroleum base, organic pesticides, is subject
to the same uncertainties as fertilizers and natural gas.

During the past five years major steps have been made
by the public research and extension agencies to develop and
demonstrate the concepts and techniques of integrated pest
management (IPM). This approach seeks a combination of pest
control agents, as opposed to relying exclusively on any one
agent, such as chemical pesticides. Chemical pesticides may
be required in IPM programs; however, they are applied only
as a last resort to keep the pests from exceeding established
threshold levels, and pesticides that pose intolerable risks
are avoided. On first impression it would seem that the
rapid and wide-scale implementation of this approach is
inevitable, because of the spectacular results that recent
efforts have produced. Compared to the chemical control
strategy, IPM is more effective, less costly, and less
hazardous to humans and the environment. However, there are
indications that the development and implementation of IPM
has moved at a snail's pace (14).

Hence, the Cornell Symposium, *Pest Control Strategies—
Understanding and Action,* has come at a very important time
in the history of pest control in this country. On one hand,
it appears that the climate is extremely healthy for marching
forth in consort to expand the concept of integrated pest
management to all sectors of pest control. A majority of
today's pest management specialists and cognizant policy
makers in government appear willing to attempt this ambitious
goal. On the other hand, there are numerous obstacles—technical, institutional, social, economical, and political—

which must first be overcome before this goal becomes a
reality. The Symposium affords a unique opportunity to
examine critically these obstacles and to weigh the benefits
of integrated pest management against the costs required to
overcome them in deciding a future course for this approach
to pest problem solving.
    Specific objectives of the Symposium are to:
    °   identify and explore alternative strategies for
          pest control
    °   discuss obstacles to more widespread implementation
          of integrated pest management practices and
          possible solutions for overcoming these obstacles
    °   provide information on alternative pest control
          strategies to extension leaders, educators,
          government officials, and agricultural leaders.
    The intent of the Symposium is to provide the background
necessary to understand various alternative pest control
strategies, hear about integrated pest management case
studies on a variety of crops and pest problems, and identify
possible action to encourage the broader use of alternative
pest control strategies.
    We are especially fortunate to have with us leading
authorities in their respective fields to present the papers
and participate in the formal discussions of the Symposium.
    However, it is up to you, the Symposium's participants—
including representatives of industry, environmental organizations, government, universities, the press, farm organizations,
and members of the public who are involved or merely interested
in integrated pest management—to determine the Symposium's
true value. We hope you will choose to participate in all
facets of the Symposium and feel free to ask questions and
to comment during the papers and discussions.

REFERENCES

 1. Metcalf, R.L. 1968. Methods of estimating effects. pp.
    17-29 *in* Research in Pesticides. C.O. Chichester, ed.
    Academic Press, New York.
 2. Smith, F.R. and R. van den Bosch. 1967. Integrated
    control. pp. 295-340 *in* Pest Control: Biological, Physical
    and Selected Chemical Methods. W.W. Kilgore and R.L. Doutt,
    eds. Academic Press, New York.
 3. Wright, J.S., R.F. Fritz, J. Haworth. 1972. Changing
    concepts of vector control in malaria eradication. Annu.
    Rev. Entomol. 17:75-102.
 4. Luckmann, W.H. and R.L. Metcalf. 1975. The pest-management
    concept. pp. 3-35 *in* Introduction to Insect Pest
    Management. R.L. Metcalf and W.H. Luckmann, eds. Wiley-

Interscience, New York.
5. Furtick, W.R. 1976. Implementing pest management programs: an international perspective. pp. 29-38 *in* Integrated Pest Management. J.L. Apple and R.F. Smith, eds. Plenum, New York.
6. Jennings, P.R. 1976. The amplification of agricultural production. Sci. Am. 235:180-96.
7. Anonymous. 1973. Where are the markets? Farm Chem. Croplife 136(1):26-30.
8. Arthur D. Little, Inc. 1976. Economic Analysis of Interim Final Effluent Guidelines for the Pesticides and Agricultural Chemicals Industry. Office of Water Planning and Standards, Environmental Protection Agency, (draft report).
9. von Rümker, R., E.W. Lawless, A.F. Meiners, K.A. Lawrence, G.L. Kelson, F. Hordy. 1974. Production, Distribution, Use and Environmental Impact Potential of Selected Pesticides. Office of Pesticide Programs, Office of Water and Hazardous Materials, Environmental Protection Agency, Washington, D.C.
10. Klingman, G.C., F.M. Ashton, L.H. Noordhoff. 1975. Weed Science: Principles and Practices. Wiley, New York.
11. Georghiou, G.P. and C.E. Taylor. 1977. Pesticide resistance as an evolutionary phenomenon. pp. 759-85 *in* Proc. XV Int. Congr. Entomol., Entomol. Soc. Am., College Park, Maryland.
12. National Academy of Sciences. 1975. Pest Control: An Assessment of Present and Alternative Technologies. Vol. 1. Contemporary Pest Control Practices and Prospects: the Report of the Executive Committee. Washington, D.C.
13. Young, J.A. and R.A. Evans. 1976. Responses of weed populations to human manipulations of the natural environment. Weed Sci. 24:186-90.
14. Flint, M.L. and R. van den Bosch. 1977. A Source Book for Integrated Pest Management. (In press).

THE ROLE OF USDA IN INTEGRATED PEST MANAGEMENT

M. Rupert Cutler

Assistant Secretary of Agriculture for
Conservation, Research, and Education
United States Department of Agriculture
Washington, D.C.

A few years ago, when life was much simpler than it is today, there were really only two subjects that were guaranteed to raise an argument: politics and religion.

Now, one is lucky to find two topics that won't create a controversy—and pesticides certainly isn't one of them.

The U.S. Department of Agriculture has been in the eye of this controversial storm since the early days of DDT and Rachel Carson's *Silent Spring*.

That's understandable because the Department of Agriculture has oversight responsibility for over 70 million acres of federal forests and 350 million acres of cropland. Both are major users of pesticides, particularly agriculture.

Chemical technology has revolutionized agriculture in the past generation.

Since World War II, U.S. farmers have sought to increase yields in relation to costs. To accomplish this, hundreds of chemicals were used to increase productivity, protect crops and decrease labor requirements.

In the beginning, at least, too little thought was given to the eventual consequences to the environment and to people. Side-effects and long-term impacts of the chemicals were unknown or ignored.

It didn't take long before the inevitable controversy occurred, and the push-and-shove between farmers and chemical suppliers and environmentalists often became bitter.

Clearly, chemicals are essential to the maintenance and increase of agricultural production. The problem is: How are farmers to use these chemicals with least adverse impact on the environment? The present solution largely resides in the regulation of the use and application of these materials. In most cases, the trade-off, so far, has pleased neither farmers nor environmentalists.

In his environmental message to the Congress, President Carter acknowledged that for several decades chemical pesticides have been the foundation of agriculture, public health, and residential pest control. He also expressed concern that of the approximately 1,400 different chemicals used in pesticide products, some, as we have begun to discover, impose an unacceptable risk to our health and our environment. To improve the safety and effectiveness of pest management, he asked the administrator of the Environmental Protection Agency to work with the Congress in enacting amendments to the Federal Insecticide, Fungicide, and Rodenticide Act that would allow the EPA to regulate directly these 1,400 active chemical ingredients, rather than the 40,000 different commercial products, which contain these chemicals in varied amounts. This would help speed the registration of safe and desirable pest control products, and it would permit swifter revocation of registration for those that pose unwarranted risks.

He also instructed the Council on Environmental Quality at the conclusion of its ongoing review of integrated pest management in the United States to recommend actions that the federal government could take to encourage the development and application of pest management techniques that would emphasize the use of natural biological controls like predators, pest-specific diseases, pest-resistant plant varieties, and hormones, relying on chemical agents only as needed.

In this message the President is reflecting the concerns of the American public. The people of this country are not only highly aware of the environment and its complexities and have a real appreciation for the benefits of diversity of species, but they also have a real concern for the impact on nontarget species, including humans, of the over one billion pounds of pesticides this country knowingly releases each year into the environment.

Dr. William D. Weil, Jr., professor and chairman of the Department of Human Development at Michigan State University, provided us with an excellent example of this dilemma in a statement he made recently before a Senate subcommittee in Washington. His testimony was related to recent reports by the Environmental Defense Fund and the Environmental Protection Agency that human breast milk increasingly contains pesticide residues and other chemical contaminants that can cause cancer and other diseases. Dr. Weil stated, "Of immediate concern to all of us is what do we say to a woman who asks: 'Should I breast feed my baby under these circumstances?'" Weil testified, "I have no easy answer. I really don't even have a good complicated answer."

The economic picture is also changing. The cost of pesticides and their application has increased dramatically in the last 10 years. The price of oil and other petroleum products is increasing while supplies are decreasing.

Because of the widespread unemployment, a large work force is available for use in places where it hasn't been in the past. If we're smart enough to utilize it, this situation gives us the opportunity to use "people-power" rather than chemicals in some of our pest management efforts, as well as the opportunity to provide meaningful jobs to people who need them.

All of these points and many others support the President's statement that now is the time to take a new look at the way we protect our crops and other resources from pests.

KEY CONSIDERATIONS OF INTEGRATED SYSTEMS

As I prepared for this presentation, I was surprised at the number of different concepts and definitions that are currently in use. One of the better definitions described integrated pest management as an approach that employs a combination of techniques to control the wide variety of potential pests that may threaten crops. It went on to say that an integrated system involves maximum reliance on natural pest population controls, along with a combination of techniques that may contribute to suppression—cultural methods, pest-specific diseases, resistant crop varieties, sterile insects, attractants, augmentation of parasites or predators, or chemical pesticides as needed. It emphasized that a pest management system is not simply biological control or the use of a single technique. Rather, it is an integrated, comprehensive approach to the use of various control methods that takes into account the role of all kinds of pests in their environment, possible interrelationship among the pests, and other factors.

One of the key factors that is omitted from most of the definitions is that integrated pest management *must* be an integral part of total management—whether it be for row crops, forest resources, or whatever. Another point that deserves special emphasis is that a truly integrated pest management system requires real multidisciplinary participation, not only in the research and development phases but also in their actual implementation on the ground.

USDA POLICY ON PEST MANAGEMENT

The Department's current policy on pest management is

incorporated in its programs for environmental quality. This
policy states that "the Department of Agriculture recognizes
that the activities associated with agriculture and forestry
and the quality of the environment cannot and should not be
separated. It is therefore the policy of this Department,
while pursuing its basic agricultural and forestry responsibilities of insuring adequate supplies of food and fiber, to
support programs which encourage favorable and responsible
relationships between man and the environment; to encourage
meaningful efforts to prevent, control and abate environmental
pollution; and to support the development of programs to enrich
the understanding, protection, and development of ecological
systems and conservation of natural resources important to
the well-being of the Nation."

Obviously, this policy statement is so broad and general
that it provides little or no guidance to USDA programs. By
September 1, I plan to develop a new policy statement that
will emphasize the use of nonchemical measures in integrated
pest management systems, as encouraged in the President's
statement. This policy will be specific and contain examples
to guide its interpretation. In developing this policy I plan
to seek the advice and counsel of anyone who would like to
give it. I've asked Mr. David E. Ketcham of my staff to
coordinate this effort. If any of you should have suggestions,
please contact Dave.

As a part of this process, I plan to review the use of
pesticides by the Forest Service who, because of the national
forests, is the Department's biggest user. This process will
begin the week of July 5.

RESEARCH, DEVELOPMENT, EDUCATION, REGULATORY, AND ACTION
PROGRAMS

Over the years the Department has strived to maintain a
balanced program on pests and on the management of pest problems in its research, development, education, regulatory, and
action programs. Our research efforts include such things as
research on pest biology and ecology; alternative methods and
systems of pest management; new use patterns of pesticides with
reduced hazard to humans and nontarget species; toxicology,
behavior, and fate of pesticides in the environment and in
exposed organisms; and economic and environmental impact of
pest management. However, in order to develop and implement
appropriate integrated pest management systems, the Department
must intensify its efforts to—
   1. accelerate the development of control tactics of all

kinds—biological, cultural, host resistance, environmental, and chemical—that preserve the ecological and genetic diversity;
2. develop improved pest detection, appraisal, prediction, and loss evaluation procedures;
3. develop simulation models and systems methodology;
4. assess and evaluate the economic benefits of the parts and whole of any integrated pest management system and make this information available;
5. accelerate the use of integrated pest management systems in extension and action programs.

Probably one of the Department's most progressive moves in recent years in the area of pest management was the Extension Integrated Pest Management Program, which was initiated in 1971. The objectives of this program are to develop and implement an effective, integrated program to prevent or mitigate losses caused by pests through the use of biological, cultural, chemical, and varietal methods of control; to develop methods for monitoring pest populations in farmers' fields; and to provide producers, consulting firms, and farmers' cooperatives with information and training in the principles of integrated pest management.

The goal of the Extension education pilot project is to teach farmers, ranchers, and homeowners how to carry out more effective pest controls; protect natural enemies; implement, where feasible, nonchemical means of controlling pests; and apply pesticides on an "as-needed" basis.

The accomplishments of this program provide us with some good illustrations of the types of benefits that we can reasonably expect from other integrated pest management programs. For example, the Extension Service found that growers in a cotton insect pest management program typically averaged two to four fewer insecticide applications than nonparticipating growers. This represented about a 35% to 50% reduction in insecticide use, and net profits of $25 to $95 per acre, depending on insect population densities. The pilot pest management projects on other commodities have shown that pesticides can be reduced 30% to 70% in situations where unwarranted or poorly-timed applications have previously occurred. Depending on the pest complex and crops involved, benefit/cost ratios of four to one to ten to one have been realized.

The Extension Integrated Pest Management Programs are cooperative programs with the involved states, and the states deserve a lion's share of the credit for making these programs go.

A number of states have demonstrated prototype predictive models to more accurately forecast local pest outbreaks and provide farmers and pest management advisors with better

decision-making capabilities. There is increasing acceptance
of the programs and a willingness of farmers to pay the cost
of monitoring field populations of pests, either by hiring
consultants or by forming grower-owned pest management cooperatives.

An outgrowth of this pilot pest management program has
been the development of many training and informational materials. States have developed educational materials to meet
their local needs. The Extension Service-USDA, working cooperatively with the states, has developed educational publications and teaching aids for diverse audiences.

Many Extension-trained professionals have entered the
private consulting business or are working for cooperatives
and the pesticide industry. The thousands of rural youth who
have been trained as pest management scouts will become better
farmers. Many are pursuing professional careers in the public
service or are working in the private sector where they are
creating profound changes in attitudes about pesticide use.

In 1973 less than 100 private consultants were offering
pest management advisory services to farmers. Today, the
number exceeds 500. Before Extension undertook this program,
fewer than a dozen service cooperatives provided any kind of
integrated pest management advice. Today, several dozen
cooperatives and farm management firms provide pest management
services. More are developing this capability as trained
professionals become available.

Many states have developed improved pest diagnostic facilities such as mobile laboratories and diagnostic clinics.
Most states conduct specialized courses on identification of
pests and provide instruction on latest recommendations to
pest control operators, aerial applicators, and pesticide
salesmen. Growers are advised on selection of resistant
varieties, cultural practices, and when to spray. Clinics
are also held in a number of major cities to inform homeowners
and home gardeners. The subject matter is interdisciplinary,
involving plant pathology, entomology, weed science, and
nematology. Extension plant pathologists have organized on a
regional basis for the purpose of reporting and forecasting
epidemics of diseases such as wheat rust and blights of corn,
tomatoes, and potatoes.

Pest control education for farmers, especially for small
farmers, can be improved in the future as more rapid communication systems are perfected and procured. Foremost on the
list will be the development of English language computer
programs that will make available information and solutions
of complex pest problems at the county office level, which
now requires consultation with Extension specialists or other
university technical personnel. Better agricultural weather

information, combined with developing new technologies for
forecasting and predicting outbreaks of pests, will improve
pest control and further advance integrated pest management
programs. Farmers will receive more sophisticated and precise
information on how to manage pests and reduce losses, thereby
increasing agricultural production and contributing to a better
environment.

The State Cooperative Extension Services will provide
training to growers, scouts, and private organizations who
offer advisory services to farmers, rather than relying on
manufacturers' representatives and salesmen of pesticides.
The State Extension Services will provide participating growers,
other farmers, home gardeners, consulting firms, and chemical
industry fieldmen with educational materials and information
on integrated pest management. Public funds will be used only
to pay professional Extension costs; develop and purchase
publications; provide program support in the form of supplies,
clerical assistance for data collection, processing, and
forecasting of pest populations; and develop specialized
communication and monitoring equipment.

An outgrowth of this program will be the creation of new
job opportunities. Thousands of youth will be provided training, summer employment, and career opportunities.

Another special effort within the Department is the Combined Forest Pest Research and Development Program. This
program is unique in the sense that this is the first time an
interagency research and development program has been adminis-
tered directly from the Office of the Secretary. I also think
that the participation under one administrative organization
of four agencies in the Department—Agricultural Research
Service, Animal and Plant Health Inspection Service, Cooperative State Research Service, and Forest Service—with state
agricultural experiment stations, universities and colleges,
state forestry organizations, and private industries represents
a milestone in federal, state, and private cooperation.

The ultimate goal of the program is to provide the necessary tools for an integrated pest management system designed
to minimize intolerable losses caused by the gypsy moth, the
Douglas-fir tussock moth, and the southern pine beetle.

After almost three years of operation, the program has
made some notable accomplishments. In August 1976 the nucleo-
polyhedrosis virus was registered by the Environmental Protection Agency for use against the Douglas-fir tussock moth.
This was the first virus to be registered for the control of
a forest insect, and it represents the culmination of 13
years of research and development. An application for the
registration of a similar virus for the control of the gypsy
moth was submitted to the Environmental Protection Agency in

December 1976. Another biological agent, *Bacillus thuringiensis,* was registered for aerial and ground application for the control of the Douglas-fir tussock moth. Excellent progress is also being made in using pheromones for the detection and evaluation of Douglas-fir tussock moth and gypsy moth populations, determining the impacts of native and introduced parasites and predators for all three insects, identifying the factors that affect forest stand susceptibility to insect attack, and developing procedures for measuring and predicting impacts.

The Gypsy Moth and Douglas-Fir Tussock Moth Programs will have completed their work and disbanded in September 1978. The Southern Pine Beetle Program is currently scheduled to complete its mission by September 1979, but it may be extended for one more year to complete, report, and implement the new technology.

During the last few months of these programs, we plan to contract with an outside firm to evaluate the Combined Forest Pest Research and Development Program and its approach to doing research and development. One key area that will be explored is the accomplishment of the accelerated program as opposed to what could have been accomplished on a business-as-usual basis. This process will begin during the summer of 1978 and be completed following the phaseout of the Southern Pine Beetle Program in September 1979 or 1980.

NEW INITIATIVES

During the past few months, the Department has taken a number of new initiatives and others are being considered. One of these is the national pesticides assessment activity, which is designed to provide maximum assistance to the Environmental Protection Agency in carrying out the process of reregistering pesticides as required by the Federal Insecticide, Fungicide, and Rodenticide Act, as amended. In this area the Department will develop thorough and objective assessments of the benefits and risks associated with the use of various pesticides. Our main interest is directed towards assisting EPA in making informed decisions so that both the needs of agriculture and the well-being of consumers will be thoroughly considered.

On April 8 Secretary Bergland established a work group on pest management. This work group is made up of representatives from the USDA agencies that are engaged in active programs directly involved or supportive of pest management. This work group is chaired by my Deputy Assistant Secretary, Dr. Jim Nielson, and is charged with the job of providing leadership and information exchange among the agencies involved. It is also responsible for taking the lead in coordinating USDA activities on pest management with those of EPA and other federal and state agencies.

The Department also sponsored a special team to study the potential uses of biological control agents. The team was composed of representatives from USDA agencies, state universities, state departments of agriculture, and the Agriculture Research Institute. This study group has just completed a report, "Biological Agents for Pest Control; Current Status and Future Prospects," which the Department will publish. This report highlights a number of opportunities for the expanded use of beneficial arthropods, nematodes, snails, microorganisms, and vertebrates for suppression of all kinds of pests. We're now considering how we might best implement the recommendations in this report in order to increase activities among all concerned interests in the research, development, and use of biological agents in the management of pest problems.

Another area that we're giving increased emphasis is the use of systems science in developing and implementing approaches to manage pest problems. The systems approach forces our scientists to consider pest management as an integral part of total crop or resource management.

The old way of doing business "down on the farm" usually involves the systematic application of pesticides at set intervals, whether they are needed or not. In order for farmers to accept and implement new techniques, they need assurance that the system will work and that it is economically sound. They know the old way works. They also know that financially they can't stand to lose a whole year's crop. Nature presents many unavoidable risks to the farmer's successful production of a crop. One way to *encourage* growers to accept the higher perceived risks of pest management systems is through an appropriate program of crop insurance that will provide adequate protection against pest losses. The Federal Crop Insurance Corporation insures most field crops on an all-risk basis in order to guarantee the producer the return of production costs. The program does not provide insurance for poor farming practices or neglect of the crop. Considerable indemnities have been paid for pests such as the pink bollworm and western corn rootworm. Congressman Jones of Tennessee has introduced legislation that would substantially broaden the scope of farm protection.

SUMMARY AND CONCLUSION

To summarize briefly, we plan to take several actions

within the next few months that will have a major influence
on the Department's pest management activities. These
include—
 1. the development by September 1 of a new policy on
    pest management that will emphasize the use of non-
    chemical measures in integrated pest management sys-
    tems; this policy will be specific and contain exam-
    ples to guide interpretation;
 2. review of the use of pesticides by the Forest Service
    to assist in the development and implementation of
    USDA policy;
 3. the utilization of the unemployed in pest management
    activities as an alternative to the use of pesticides,
    where possible;
 4. the expansion of the Extension Service's integrated
    pest management programs to make pest management
    technology available to more farmers, ranchers, and
    homeowners;
 5. the evaluation of the use of accelerated research and
    development programs to provide pest management tech-
    nology for critical problems within a short period of
    time;
 6. the full support of our continuing research programs,
    which are the backbone of our research and development
    effort;
 7. the expanded use of biological control agents in those
    areas where possible and appropriate.
   So, as you can see, the USDA has a major role in integrated pest management. However, this role is broader than our
research, development, technology transfer, and action programs.
It also includes the obligation of providing responsible leadership in the use of integrated strategies in all areas of
agriculture and natural resources. This is an obligation we
intend to fulfill.
   Another obligation that we see is that of working closely
and cooperatively with the Environmental Protection Agency,
other agencies, states, industries, groups, and associations
to accomplish our mutual goals. I mention EPA in particular
because of our closely related interests in the pest management
field. If we are both to redeem our respective responsibilities,
we must work closely together as a team rather than as adver-
saries. If we have real differences of opinion on particular
issues, and I'm sure we will have, then we're going to sit down
together and work them out. We must keep the lines of communication open, and show the people of this country that big
government can not only be sensitive to their needs but that it
can also be responsive.
   We're looking forward to working with each and every one
of you, and welcome your advice and counsel.

DISCUSSION

B. DAY: Do you propose an expanded budget for this kind of research?

M.R. CUTLER: The new zero-based budgeting allows restructuring the budget of each of our agencies. We are making a line-item of IPM, and giving it high priority.

T. CLARKE: I'm concerned about applied and primary research. How do you integrate land-grant and private universities with ARS in developing IPM programs?

M.R. CUTLER: Much of the research should be done at universities. Nationally, ARS is responsible for basic research, but often the problems are local ones. Also, multidisciplinary teams are perhaps better assembled at universities, although ARS is becoming more multidisciplinary. Much IPM research should be done at land-grant and other universities, and at agricultural stations.

J. COX: How long will it be until county agents implement integrated pest management programs?

E. SMITH: Where pest management programs are ready for implementation, they already are being used by county agents. For example, we have an apple pest management program in western New York State that has been implemented.

UNIDENTIFIED: The chemical companies are such a massive force, will they hinder the adoption of IPM programs?

M.R. CUTLER: This could be a problem; for instance, chemical companies could influence Congressional appropriations. I am optimistic, however, that with the problems of pesticide use, IPM will carry the day.

N. CHANDLER: As a farmer, I'm concerned about how to attract the extra labor required for alternative pest control programs.

M.R. CUTLER: On our forest lands, we're looking for alternatives to the use of 2,4-D and 2,4,5-T. We're considering using a CCC-type of job corps. We might be able to expand the role of the young adult conservation corps to include private as well as public lands.

B. DAY: Is it your policy, then, to oppose the use of 2,4-D and 2,4,5-T on forest lands?

M.R. CUTLER: We are not going to take a position on that now, but we are putting it under review. 2,4,5-T and the dioxins that are often associated with it make us uncomfortable, and we certainly will avoid using these herbicides if possible.

UNIDENTIFIED: Do your future IPM plans include the boll weevil eradication project?

M.R. CUTLER: APHIS is not one of my agencies! I think this program demands some scrutiny.

R. HELGESEN: Do you plan to increase extramural funding from the USDA to universities?

M.R. CUTLER: By FY 1979, we plan to have competitive grants for IPM research at an expanded level; these will probably be concentrated in a few well-qualified institutions.

H. OLKOWSKI: The weed management question interests me, because in the San Francisco Bay area we were approached by a union of park workers who wanted to avoid using pesticides, and were looking for alternatives. Here's a union that *wants* to substitute labor for chemicals. This might have wider application, especially with farm workers concerned about exposure to pesticides.

UNIDENTIFIED: Does the USDA support resident instruction?

M.R. CUTLER: Education *per se* is HEW's responsibility. We do support it indirectly by providing facilities and so on.

Part II
Complexity of Pest Management

INTEGRATED PEST MANAGEMENT—A BIOLOGICAL VIEWPOINT

Richard D. O'Brien

Division of Biological Sciences
Cornell University
Ithaca, New York

A conference of this kind is like a banquet. You come
here with great expectations of the things you will ingest,
you fill yourself with great quantities of mind-food, so that
by the time the affair draws to a close you can hardly imagine
getting any more down, and if all goes well you remember it
for years in the future. Now when many people sit down to an
ordinary meal, they start right in with the meat and potatoes.
But at a banquet that would be unthinkable. Instead, you
start with a few drinks, and before you get to the main course
you have what the French call hors d'oeuvres, meaning out of
the main work; or what the Italians call antipasto, meaning
before the spaghetti. The names clearly show that this little
course is nothing more than a preparation for the real thing,
and that is exactly what my talk will be about. Like an antipasto, it is designed to stimulate the appetite, but not to
fulfill it. It will be made up of an olive here, an anchovy
there, a touch of salami, and a piece of ham. Do not mistake
this for the main course, which will be started by Ray Smith.

In the spirit of being a good antipasto, I would like to
present several quite different items to you, and each one
will take the form of a theme which I hope and believe represents a feature of contemporary biological thinking, which has
relevance to the question of integrated pest management. My
first theme is:

The second law of thermodynamics is still alive and well.

One way of stating the Second Law is "spontaneous processes tend toward states of lower energy and greater disorder." A corollary of this is that if one wants to push a
system in a way that maintains a high degree of order, one

needs a large amount of energy to do so. Now our whole agriculture is based upon orderliness. Both in the plant and
animal realm, we invariably seek a local monoculture that is
absolutely unlike the tremendous diversity characterizing
spontaneous living systems. That these monocultures bring
upon themselves plagues of insects, fungi, and weeds is not an
accident. It is an expression of the second law of thermodynamics.

Many years ago I was visiting England at the time of the
Suez crisis. My mentor there was Sir Rudolph Peters, a
distinguished biochemist of the old school. Sir Rudolph's
analysis of the Suez crisis was as follows "Everything is
going along perfectly splendidly, and then some jackass over
in Cairo or Timbuktu or somewhere makes a fool of himself, and
we are all put to the most frightful inconvenience." The
analagous thinking in the pest control area is to imagine that
one's crop is getting on perfectly splendidly and then some
fearful disorder breaks out in which insects or fungi or weeds
invade it. Both views are of course completely misguided. In
both cases, whether it be British foreign policy before Suez,
or a good stand of oats without weeds, one had established a
very transitory and very unstable and completely artificial
situation, which was absolutely bound to deteriorate sharply.

In a similar way, the older view towards pest control saw
each pest as an invader into a harmonious and stable system,
and of course the invader needed to be utterly thrown back in
a way as dramatic and total as that in which it seemed to
appear out of nowhere. By contrast, a contemporary view would
be to regard the total system as an elaborate set of equilibria, which need to be understood and regulated to the best
possible extent ahead of time, and adjusted by appropriate
means when one component of the equilibrium appears to be
gaining a disproportionate hold. The strategy that accom-
panies this is not one of relaxation until the foe strikes,
but rather an understanding that desirable and undesirable
factors are invariably present simultaneously at all times,
and that one needs to quantify this balance on a continuing
basis, and move to redress imbalances long before they achieve
crisis proportions.

There is another corollary to this view of the situation.
It is that with any monoculture, and therefore with virtually
all of agriculture in any particular year, one has already a
highly artificial and unstable situation. This is perhaps not
adequately recognized by those who call for "natural" or
"organic" or "truly biological" approaches to pest control.
The maintenance of the thoroughly artificial system that our
society must have in order to survive may often require
artificial remedies. Their evaluation should not rest upon

their degree of artificiality, but upon their ability to maintain the strange balance that we demand in our system.

Let me turn now to a second theme:

Biological systems are always complex and interacting.

During my aforementioned return to visit the country of my youth, I talked with Professor Malcolm Dixon of Cambridge, and discussed the relationship of enzymes to the cell and its organization. Dixon's view was simple. "A cell is a bag of enzymes," he said. "If you understand each one of the enzymes completely, you will understand how the cell operates." That was in 1957. Almost the whole of the intervening twenty years has been made up of a growth of understanding that exactly the opposite is true, and that the most interesting feature of a cell is the way that it regulates itself, by a series of interactions between its different parts. Study of any one individual component enzyme tells one remarkably little about that regulatory feature of the cell's life.

I believe it is now widely recognized that a similar situation holds in the relations between organisms. The earlier approach was to study each organism in relative isolation, and if one of the organisms, such as a particularly predatory insect, became a nuisance, one applied one's understanding of that individual to attempts to squelch it. It reminds me of an old joke in the British humor magazine "Punch". Two rough-looking men, A and B, see a man walking down the road. A says to B, "Who's 'e, Bill?" Replies B, "I dunno. A stranger." "Well, then," says A "'eave 'alf a brick at 'im."

The fact is that the pest that suddenly appears to invade the crop is often no stranger. It or its forebears were there all along, and conditions have now become favorable for it to become a dominant feature of the biological scene. What we need is understanding of the factors that maintained it at an acceptable level in the first place, and the factors that permitted it to achieve outbreak proportions subsequently. In essence, this is a problem in population dynamics, but it must be clearly focused not only upon that individual species, but on the many species that share the uneasy equilibrium characterizing a biological system at any one moment. It is unfortunate that there is an extremely imperfect understanding of the factors governing these equilibria between insects and plants (for instance) in spite of decades of study and arguments. A number of mathematical models have been developed that are reasonably effective in describing empirically the fluctuations of populations in purely descriptive terms, but what is needed is a complete understanding of the biological factors behind those variations, so that one can determine

which factors can be successfully manipulated to achieve the
desired outcome. Until this fundamental biological problem is
solved, pest control will be almost totally empirical in the
years ahead as it has been in the past. I am entirely unqualified to review the validity of the arguments in the field
of population dynamics, but I would like to point very briefly
to the considerable depth of knowledge that exists with
respect to the corresponding equilibria within cellular
systems, to illustrate the depth of understanding necessary at
the population level if we are to be equipped for our task. I
shall point to three examples, with the intent of demonstrating how complicated but ingenious they are, and how regulation can be achieved through radically different mechanisms.
We should certainly not expect to find a simpler situation
when we turn to the regulation of populations of organisms
rather than populations of molecules.

The first of the three completely different mechanisms is
that of repression, which is extremely common in bacterial
systems, and of which a comparable version may exist in higher
organisms. It has been particularly well worked out with a
bacterial cell that contains the whole machinery to synthesize
lactose-metabolizing enzymes, but which does not normally
synthesize these enzymes unless considerable amounts of
lactose suddenly put in an appearance. Figure 1 shows dia-
grammatically the portion of the bacterial DNA involved, with
the diagram at the top indicating the normal situation when
lactose is absent, and the cell is not producing the enzymes
necessary for its metabolism. The genes for synthesis of
lactose-metabolizing enzymes are symbolized as Z and Y and A,
and if the polymerase enzyme that is attached to the promoter
region could only get going, it would zip along the DNA, and
cause genes Z and Y and A to become active and to induce the
formation of the messenger RNA, which would subsequently cause
the synthesis of the enzymes. In the normal situation, the
polymerase is blocked by a protein called a repressor, which
gets in the way of the action of the polymerase. But if
lactose should suddenly appear in quantity, it has the ability
to combine with the repressor to convert it to an inactive
form, which no longer has an affinity for the DNA; consequently it diffuses off and permits the polymerase to get to work,
and is shown in the bottom half of the figure. This, then, is
the repression and derepression mechanism of the control of
production of a whole package of enzyme molecules.

A quite different mechanism is involved in the regulation
of the important molecule ATP, which is the immediate source
of energy for a whole variety of biological reactions, including muscle contraction and nerve transmission. This is

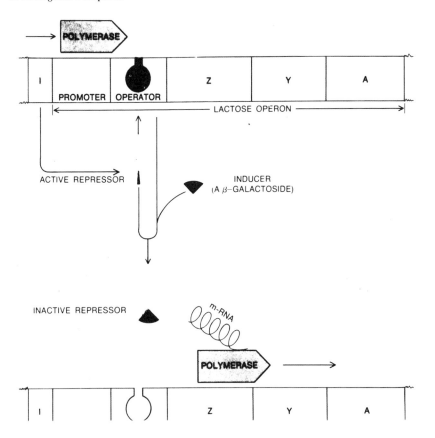

Fig. 1. The lactose operon. Top: The normal situation, in which genes Z, Y, and A are not expressed. Bottom: When lactose accumulates, the repressor is removed and the genes become expressed [from (1)].

the *tight coupling* mechanism. Figure 2 shows the complicated pathway which is necessary for the simultaneous oxidation of foodstuff, typified by pyruvate on the far left, and the coupling of this oxidation to the production of ATP from its precursor ADP. We do not have to worry about the details at all, but only to indicate that if ADP is not available at the three locations shown at sites I and II and III, then the machinery grinds to a halt, and no more ATP is produced. Then when the body begins to use the ATP that has been produced, it does this by converting that ATP to ADP, and this automatically starts the machinery going again. Thus, the oxidative part of the pathway at the top of the figure is tightly coupled to the phosphorylating steps at the bottom, and the

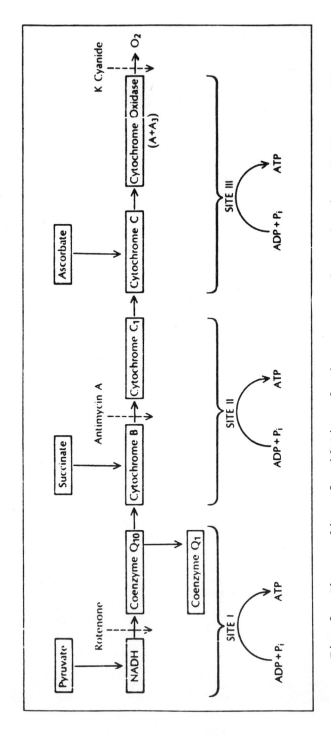

Fig. 2. The coupling of oxidation of substrates to the production of ATP from ADP [from (2)].

28

body has the ability to turn off ATP production and save its resources for a future need.

A third control mechanism, again entirely different, is extremely common; we can call it end product inhibition. A beautiful example is shown in figure 3. The compound CTP at the bottom of the figure is synthesized through a whole series of steps starting at aspartic acid. One of the key enzymes in this long pathway is aspartic transcarbamylase, shown in the figure as ATCase. This enzyme has built into it a special

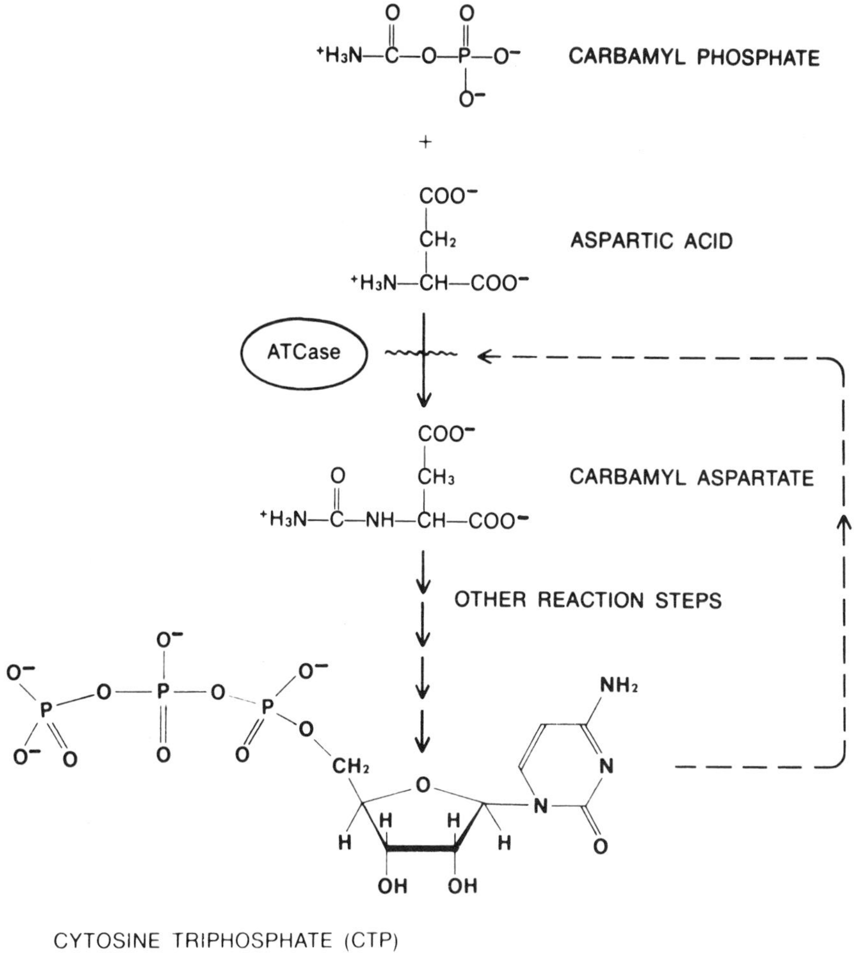

CARBAMYL PHOSPHATE

ASPARTIC ACID

ATCase

CARBAMYL ASPARTATE

OTHER REACTION STEPS

CYTOSINE TRIPHOSPHATE (CTP)

Fig. 3. End product inhibition (also called feedback inhibition). Overproduction of CTP causes it to bind to ATCase (aspartic transcarbamylase) which inhibits the enzyme, turning off CTP production [from (1)].

area that can attach to CTP, and when such an attachment
occurs, it turns the enzyme off. Thus when the pathway gets
going, the ATCase may be going at full speed; but when it
achieves its objective of piling up enough CTP to provide for
the needs of the cell, the CTP turns off the pathway by closing
down the ATCase. Thus the end product of the reaction inhibits
the first step in the reaction.

I promise not to show any more nonpesticidal biochemistry, but I hope to have proved the point that at the cellular level the way in which a population of molecules organizes
itself is complicated and ingenious, and can operate through
very different mechanisms. Things will be just as complicated
when we finally understand just how it is that a plant re-
covers from a moderate defoliation by growing new leaves, or
how the population of a plant-eating insect somehow escapes
from control by its own parasites and predators to achieve
outbreak proportions. Unless we understand the mechanisms
underlying these regulations at the whole-organism level, and
in the same sort of ingenious detail that we have just seen
for molecular mechanisms, we will not understand the extent to
which we dare interfere with the complicated equilibria of
plants and their pests, and the parasites and predators that
keep the pests, under many conditions, at an acceptable level.

My third biological theme is:

The interactions of most systems are controlled by numerous
factors.

This theme sounds like a statement of the obvious. But
I believe it is not truly obvious, and that a departure from
monofactorial thinking to polyfactorial thinking has characterized the growth of biology in recent decades, and its
application to pest control is absolutely crucial.

Figure 4 shows a diagram of a series of steps, which
could be chemical or physical or biological, in which A leads
to B, and then B to C, and so on, with the velocity of each
step characterized by a rate constant k. Such a simple chain
of steps is called a catenary series, and the simple law governing it is that the rate at which the whole chain proceeds is
governed by the rate of the slowest step. If, for instance,
the k_2 step is the slowest, then the rate at which A is con-
verted to E is entirely governed by the k_2 step. If one were
trying to interfere with the overall conversion of A through
E (for instance, if this is a series of biological events that
led to the outbreak of a pest population at E) then it would
do relatively little good to try to reduce the rate of k_1 or
k_3 or k_4, because they are not the rate-limiting steps. But
reduction of the k_2 step by 50% would automatically reduce

A CATENARY (· = CHAIN-LIKE) SERIES.

$$A \xrightarrow{k_1} B \xrightarrow{k_2} C \xrightarrow{k_3} D \xrightarrow{k_4} E$$

The slowest step determines the rate of the overall reaction $A \longrightarrow E$.

Fig. 4. A catenary series.

the overall conversion of A to E by 50%. This is an attractive and simple system, and many chemical examples exist. As far as control is concerned, we have here a monofactorial system in which manipulation of a single factor, in this case the k_2 step, would be adequate to manipulate the whole system.

A great deal of laboratory thinking rose out of this kind of monofactorial approach. The classic way of doing an experiment is to control all of the variables except one, and then to manipulate that one remaining variable and examine the consequences. For instance, in exploring the best fertilizer for growing tomatoes, we might provide optimal light, heat, and water to a group of tomatoes, and then vary the fertilizer mix and observe the effect upon yield. It is relatively recent for experimentalists in laboratories to understand that this monofactorial approach is sometimes of limited value. By contrast, the agronomists have always tackled things from the polyfactorial viewpoint, because they are seldom able to control the variables even if they wanted to. Thus, a corn experimentalist will vary the fertilizer mix applied to several corn plots, but he is well aware that different corn plots will vary in their soil and micro-climate and other factors, so he will finish up by conducting an analysis of variance in order to sort out the growth improvements caused by each fertilizer, and separate it from the variations in yield that were related to those other factors.

Let me give you a completely different sort of example of polyfactorial thinking, drawn from the field of animal behavior. Biologists have been interested in determining how a homing pigeon, released in strange country a hundred miles from home, succeeds in getting back to his starting place. The question used to be couched in the form, "Does the pigeon utilize the sun as a compass, or does he use the earth's magnetic field, or does he use visual landmark identification?" One way of studying the question of the use of the earth's

magnetic field was to attach magnets to the pigeon and see
whether that interfered with its navigation. Several people
performed such experiments and were able to show quite plainly
that the magnets did not interfere with the bird's performance.
But William Keeton at Cornell undertook to reexamine this
question in a more sophisticated way. He had already demonstrated that the normal pigeon under normal conditions did indeed make use of the sun as a compass. Because the sun's
position shifts throughout the day, anyone who uses it as a
compass needs a clock in order to compensate for this shift,
and the pigeon contains just such a time-compensated sun compass. One can demonstrate this persuasively by fouling up the
pigeon's internal clock by bringing it up in a room with an
artificial and incorrect light sequence in which one makes the
sun come up at midnight; and these modified pigeons will now
make a consistent and calculable error in their navigation
when they are released under real conditions. Thus, if their
clock is shifted by a quarter of a day, they will make an
error of almost precisely 90°. But does this prove that they
are quite unable to make use of the earth's magnetic field?
Keeton proceeded to study this by training pigeons to fly
under conditions of overcast that were so complete as to block
out the sun. Because pigeons do not like to fly under these
conditions, past experimenters had done most of their experiments on days with little overcast. Keeton was able to show
that although pigeons flying on sunny days were not disoriented by having magnets attached to them, pigeons flying under
overcast conditions were severely disoriented by the magnets.
It was evident that the pigeons would prefer to use the sun
compass when it was available, but if it was unavailable they
would then fall back upon some form of magnetic compass (3).
This work has been expanded considerably, and it is likely
that pigeons have four or five quite different navigating
devices, which can be used under appropriate circumstances,
but which are not used when the preferred sun compass is available. This turns out to be a rather general feature of animal
behavior, in which there is extensive redundancy, with the
result that if one system fails, there are others to fall back
upon.

Monofactorial thinking shows up in every kind of science.
My own research deals in large part with how toxic agents
interact with their targets. The commonest insecticides in
use today are organophosphates and carbamates, and the target
for these insecticides is the enzyme acetylcholinesterase.
Now it is generally recognized that one can account for the
way in which the ordinary substrate for this enzyme, i.e.,
acetylcholine, reacts with acetylcholinesterase at a single
binding site; after the binding, the substrate reacts in a

second step with a catalytic site. It was established many
years ago that organophosphates and carbamates go through a
very similar set of reactions, and they undoubtedly react with
the same catalytic site as does acetylcholine. For that
reason, it was natural to imagine that the inhibitors also
bound to the same binding site. For years, in spite of the
tremendous variety of organophosphates and carbamates that
were invented, attempts were made to squeeze the data to make
them fit the thought that everything was binding to the same
site. But in recent years it has become clear that one simply
cannot account for all of the data in this way, and one is
forced to recognize the existence of numerous binding sites
grouped around the catalytic site, as shown in figure 5 (4).

From these two completely different examples, the conclusion I want to stress is that in the real world one has to
recognize the interplay of numerous factors, and one neglects
this fact to one's peril. In a similar way, when a pest population becomes of unacceptable size, it does so because of
the interplay of several competing factors. The assumption
that the pests simply grow too numerous, and that the solution
is to knock them down to near zero by quick chemical control,
may be an invitation to disaster, particularly when the chemical control interferes with the interaction between host and
pest and the pests' parasites and predators in an uncontrolled
way. It is therefore essential to study the variety of interactions that may occur and the way in which a particular
treatment may influence each of these variables. It is the
integration of these various levels of understanding, as much
as the integration of a variety of kinds of treatments, which
lies behind integrated pest management.

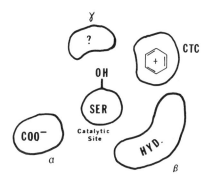

Fig. 5. A speculative map of the active surface of
acetylcholinesterase.

My fourth theme is:

To understand the interaction of organisms with exotic
compounds, such as pesticides, one needs to understand the
interaction of the organism with its environment.

In 1961, Hal Gordon of Berkeley observed (5) that those
species of insect larvae that fed on a variety of plants were
tolerant to many insecticides, and suggested that this tolerance might be related to the fact that such insects needed to
endure biochemical stresses because of the diversity of their
foods. This possibility has been explored in much more detail
in recent years. It is apparent that such insects contain a
kind of multipurpose enzymic degradation system called mixed-
function oxidase (MFO). This plays a crucial role in the
degradation of insecticides applied to the insect, and an
individual with a high enough level of MFO can survive a dose
that would otherwise be lethal. Two things have recently become apparent. One is that the MFO plays the normal function
of destroying secondary plant substances, which the plants
probably put there in the first instance to make themselves
unattractive to insect predators. It evolved in a response to
those miserable plants that wanted not to be eaten, rather
than because of the miserable human beings who were trying to
slay the insect with insecticides. The second fascinating
feature is that the MFO is inducible; that is to say, that an
individual may normally have a fairly low level, but when an
appropriate secondary plant substance (or for that matter, a
pesticide) comes into the diet, then the insect increases its
level of MFO in order to tackle this new problem. This induction is, of course, closely parallel to the earlier case I
described of lactose metabolism, in which the necessary
enzymes were not produced until the lactose presented itself.
Figure 6 comes from the recent work of Brattsten, Wilkinson,
and Eisner (6), and demonstrates how naturally occurring
secondary plant substances such as alpha-pinene can cause the
MFO level in the armyworm to triple. Figure 7 shows that this
reaction is quite rapid, and that within a few hours of a diet
containing alpha-pinene (top line) the level of MFO has risen
dramatically. You can see that the insect has developed a
strategy for degrading poisonous compounds, which no doubt
predated the invention of synthetic insecticides by millions
of years. It is ironic that some of the components of insecticide mixes that we thought were inert, primarily the solvents in which these compounds are suspended, turn out to be
very successful agents for inducing MFO, as Table 1 shows, and
in this way promote the insect's ability to degrade the pesticide in which the solvent is applied (7). Thus the solvent,

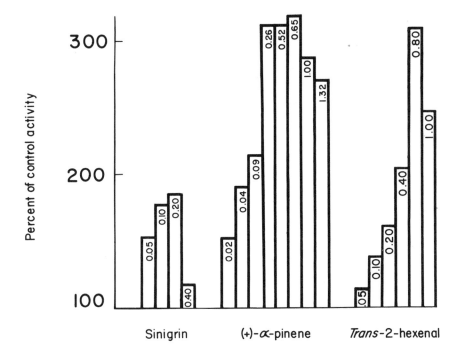

Fig. 6. Effect of three secondary plant substances on mixed-function oxidase of armyworm midgut, after 24 hr. Numbers show % of substance in diet [from (6); copyright 1977 by the American Association for the Advancement of Science].

which for years we have treated as a purely physical component in the system, to be used in a way that most promotes the distribution of the pesticide, turns out biologically to be highly significant.

My fifth and final theme is:

Organisms have to be considered in their entirety. The molecular, physiological, developmental, and behavioral aspects are all facets of the one organism.

To illustrate this big theme with but one example, the nervous system (which is the target for most insecticides) is intimately linked with the behavior of the insect (which is the target for pheromonal control). The behavior is an expression of, and is limited by, the nervous system. And the nervous system is in large part designed to program that behavior.

From this perspective, all chemical agents are simply

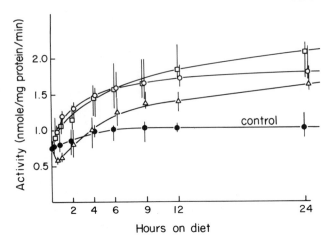

Fig. 7. *Rate of increase of mixed-function oxidase in armyworm gut after feeding (top to bottom), 0.04% α-pinene, 0.1% sinigrin or 0.2% trans-2-hexenal [from (6); copyright 1977 by the American Association for the Advancement of Science].*

interferers with one or another component of the behavioral-physiological-biochemical complex. There is no fundamental difference between a pheromone, which applied in excessive amounts causes a behavioral disorientation, and a mimic of chemical transmission such as nicotine, which applied in sufficient amounts causes disorientation in the central nervous system. And the organophosphates and carbamates are in no way different from the analogs of juvenile hormones. In both cases they are foreign compounds that disturb the normal information processing; in one case, the central nervous system is involved, in the other the developmental system. Let me stress that I am not simply trying to score semantic debating points. This discussion is only useful if it establishes that stressing the similarities of these chemical approaches to interfering with an organism has some implications for the strategy to be used in developing new and more useful ones. Of course, there are common problems of formulation, application, penetration, and metabolism. But it goes further than that. In all these cases the biological and chemical developments need to go hand in hand, and this is not easily achieved. Most synthetic programs are industrial, and most basic biological research is in universities and government. We need imaginative new ways to fuse these together, and to promote exciting interactions of biologists and chemists.

Table 1. Effect of Insecticide Solvents on the *N*-demethylation of *p*-chloro *N*-methylaniline in Midgut Tissues of the Southern Armyworm[a]

Compound	Percent in diet	Specific activity (nmole mg^{-1} min^{-1})	Percent of control activity
Phenobarbital	0.25	2.96 ± 0.18	305
Benzene	0.20	0.99 ± 0.11	102
Naphthalene	0.20	1.45 ± 0.12	149
1-Methylnaphthalene	0.20	3.35 ± 0.20	346
2-Ethylnaphthalene	0.20	3.29 ± 0.25	340
Solvesso Xylene	0.20	2.21 ± 0.17	228
HAN	0.20	3.69 ± 0.27	380
Amsco-Solv E-98	0.20	4.35 ± 0.30	449
Pansol AN-2	0.05	2.58 ± 0.20	267
	0.10	2.75 ± 0.21	283
	0.20	2.74 ± 0.21	282
Pansol AN-2K	0.05	2.58 ± 0.19	267
	0.10	3.34 ± 0.22	345
	0.20	3.15 ± 0.23	325
Mentor 28	0.20	3.87 ± 0.28	399
Hess odorless spray base	0.20	2.51 ± 0.20	259
Diisooctylphthalate	0.20	2.07 ± 0.20	214
Benzoflex 9-88	0.20	1.49 ± 0.15	154
Amsco odorless spray base	0.20	1.25 ± 0.10	129
60-second spray oil	0.20	1.21 ± 0.12	125
70-second spray oil	0.20	0.99 ± 0.10	102
100-second spray oil	0.20	0.99 ± 0.11	102

[a]Newly molted sixth-instar armyworms were given free access for 24 hours to semidefined artificial diets containing the compounds. Armyworms of the same age fed diets with no added chemicals were used as controls. The *N*-demethylase specific activity in midguts from control armyworms was 0.97± 0.10 nmole per milligram of protein per milligram of protein per minute (± S. E.). Experiments were replicated two or three times [from (7)].

Now the antipasto is almost finished, and you are ready for the meal itself. I have said relatively little about integrated pest management directly. But I hope to have traced out the way in which, in a variety of biological spheres, our thinking has evolved from a relatively simplistic

status in which limited knowledge can provide a perfectly
adequate basis for practical action, to one in which one needs
to recognize not only the identity and properties of each of
the components of the ecosystem under study, but the nature of
the relation between them and what governs their respective
roles. This symposium bears the subtitle "Understanding and
Action." In terms of understanding, I think a consequence of
what I have said is that one needs to understand far more
about organisms and their interactions than we have conceived
in the past, and that this understanding must extend to biochemical, physiological, behavioral, and population levels,
and above all to the interactions between these various
levels, if we are to have the knowledge that provides the
basis for effective action.

REFERENCES

1. Wilson, E.O., T. Eisner, W.R. Briggs, R.E. Dickerson, R.L.
   Metzenberg, R.D. O'Brien, M. Susman, W.E. Boggs. 1973.
   Life on Earth. Sinauer, Stamford, Connecticut.
2. Racker, E. 1974. Inner mitochondrial membranes: basic and
   applied aspects. Hospital Practice, February:87-94.
3. Keeton, W.T. 1974. The mystery of pigeon homing. Sci. Am.
   231:96-107.
4. O'Brien, R.D. 1976. Acetylcholinesterase and its inhibition. pp. 271-96 *in* Insecticide Biochemistry and Physiology. C.F. Wilkinson, ed. Plenum, New York.
5. Gordon, H.T. 1961. Nutritional factors in insect resistance to chemicals. Annu. Rev. Entomol. 6:27-54.
6. Brattsten, L.B., C.F. Wilkinson, T. Eisner. 1977. Herbi-
   vore-plant interactions: mixed-function oxidases and secondary plant substances. Science 196:1349-52.
7. Brattsten, L.B. and C.F. Wilkinson. 1977. Insecticide
   solvents: interference with pesticide action. Science 196:
   1211-13.

DISCUSSION

UNIDENTIFIED: Regarding the different mechanisms that pigeons
use for orienting themselves, there are still many unknowns in
the homing process. Smell and sound may be important; a simple
compass is insufficient.
R.D. O'BRIEN: That's true. You need a map as well as a

compass, and we have no idea what constitutes the pigeons' "map."

H. OLKOWSKI: There are psychological problems, and problems with the English language, in interpreting integrated pest management programs for farmers' use. It is difficult to explain these programs in understandable terms, while maintaining their complexity. It was much easier to say, if DDT works, why don't we use it?

R.D. O'BRIEN: Yes, it's easy to say spray this and it will disappear. IPM will reinforce the discrepancy between producer and professional considerations; these groups used to be more on the same wavelength.

J. COX: What are the effects of pesticides on soil microorganisms?

R.D. O'BRIEN: Here the mystery is even deeper; interaction is the important term. These interactions are not well understood.

UNIDENTIFIED: Can you anticipate the side-effects of pheromones and juvenile hormones?

R.D. O'BRIEN: The effects of pheromones are probably minimal, especially to vertebrates. They may affect nontarget insects, but this is unlikely, because they are very specific in nature. Juvenile hormone analogues include many epoxides and alkylating agents. Almost certainly some of these will be found to be carcinogenic.

HISTORY AND COMPLEXITY OF INTEGRATED PEST MANAGEMENT

Ray F. Smith

Department of Entomological Sciences
University of California
Berkeley, California

In the brochure that announced this Symposium on Pest Control Strategies, both the terms "integrated pest control" and "integrated pest management" were used. I could not detect any clear distinction between them as used in the brochure. As both are relatively new terms that have been defined and used in several different ways, I believe it will be useful to comment briefly on definitions. In California, it was the conflict between the entomological group that wanted to "control," i.e., kill insects with chemicals, and the competing group that wanted to "regulate" insects with parasites and predators that spawned in the mid-fifties the first use (1,2) of the term "integrated pest control." This first use emphasized the integration of the two tactics of biological control and chemical control into a pest management system and stressed the point that both are clearly essential to efficient and economical pest control. Subsequently, the Food and Agriculture Organization of the United Nations (FAO) adopted the term "integrated pest control" in the early sixties, held an international conference on the topic in 1965, and in the following year established an FAO Panel of Experts on Integrated Pest Control. At that time, there were a number of competing terms in use meaning essentially the same thing, e.g., "harmonious control," "pest management," "rational control" and "modified spray program." The FAO Panel supported the use of a single term "integrated pest control" and in its deliberations on the topic emphasized the importance of integrating all compatible tactics into a pest control strategy.

The definition of integrated pest control adopted by the FAO Panel more than 10 years ago was a very broad one. I offer to you here a similar broad definition. Integrated pest

41

control is a multidisciplinary, ecological approach to the
management of pest populations, which utilizes a variety of
control tactics compatibly in a single coordinated pest
management system. In its operation, integrated pest
control is a multi-tactical approach that encourages the
fullest use of natural mortality factors complemented
when necessary by artificial means of pest management. Also
implicit in its definition is the understanding that imposed
artificial control measures, notably conventional pesticides,
should be used only where economic injury thresholds would
otherwise be exceeded. As a corollary to this, integrated
pest control is not dependent on any single control
procedure or tactic. For each situation, the strategy is to
coordinate the relevant tactics with the natural regulating
and limiting elements of the environment.

Following FAO's early lead and that of the International
Organization of Biological Control (IOBC) in Europe, there
has been rather uniform adoption of the term "integrated
pest control" in plant protection circles throughout the
world since 1966 (with the exception of the United States).
With the publication of the Council on Environmental Quality
document entitled, "Integrated Pest Management," in 1972,
there has been widespread use in the United States,
particularly in federal circles, of the term "integrated pest
management." As near as I can determine from current usage,
integrated pest management has precisely the same meaning as
integrated pest control. They are absolute synonyms, so it
is not essential that a choice be made between them. Which-
ever term one may choose to use, it is far more important to
understand the fundamental ecological approach that underlies
both.

EVOLUTION OF PEST CONTROL PRACTICES

The history of integrated pest control goes back much
farther than the mere coining of the term in the early
fifties and the subsequent elaboration and clarification of
the concept. The origins are deeply rooted in the evolution
of pest control practices as developed by entomologists and
plant pathologists in the nineteenth century.

Human history is a series of attempts to gain increasing
control over the environment. At first this control was
minimal to the degree that poor shelter and unstable food
supplies imposed severe population constraints. The gradual
gain in man's capacity to control his environment parallels
the gradual rise of civilization. But as man aggregated into
villages near rivers and planted crops nearby, he encountered

increasingly severe attacks by pests on the crops. (Pests, an anthropocentric concept, is used here as a broad generic term for all forms of life that cause damage to crops and other things of value to humans or that in other ways annoy them.) For thousands of years, man could do nothing about these pests but appeal to the power of magic and a variety of gods. For the most part, early humans had to live with and tolerate the ravages of plant diseases and insects, but gradually they learned how to improve their condition through "trial and error" experiences. These improvements included the beginnings of pest control or pest management, the preferred term today.

Prior to the emergence of the crop protection sciences and even before the broad outlines of the biology of pests and the causal nature of plant diseases were understood, humans evolved many cultural and physical control practices for protection of their crops. Many of these practices were subsequently proven to be scientifically valid even though originally they were derived mostly through empirical methods. These pest control practices, still valid and useful today, include sanitation (i.e., destruction or utilization of crop refuse, roguing of diseased plants, etc.); tillage to destroy overwintering pests and disease inoculum; removal of alternate hosts of pests; rotation of crops to discourage buildup of damaging populations of weeds, insects, and other pests; timing of planting dates to avoid high-damage prone periods; use of pest-free seed and seedling methods; use of trap crops; selection of planting sites; pruning and defoliation; isolation from other crops; and management of water and fertilizers (3).

The appropriate use of these cultural methods reduced the damage potential to crops of essentially all pests and provided satisfactory economic control of some. But there were many pests of high damage potential that could not be controlled adequately by early agriculturists with any combination of known cultural control methods. As biological knowledge grew during the eighteenth and nineteenth centuries and as pest problems became more severe because of an intensification of agriculture and the introduction of pests into new areas, man became increasingly preoccupied with the search for more effective pest control measures. The discovery during the 1850s that fungi can produce diseases of plants opened the way for the scientific study of agents to control diseases, and the principal search was for chemical compounds (4). Various chemicals and concoctions were recommended for the control of insects and diseases as early as the eighteenth century (5). Much progress was made in the technique of chemical control of both plant diseases and

insects during the last quarter of the nineteenth century.
As early as the turn of the twentieth century, there were
optimistic expectations that both diseases and insects
would ultimately be controlled by chemical pesticides alone
(6).

Plants resistant to insects and diseases were recognized
in the nineteenth century, but the deliberate development of
pest-resistant crop varieties was not possible until after
the rediscovery of Mendel's laws of heredity in 1900.
Following this breakthrough, the approach was quickly
exploited for the control of important plant diseases of
cereal and other crops, but was pursued less vigorously for
insect control until quite recently.

The success with chemical control and host resistance
in controlling plant pests detracted from the importance of
cultural control in many instances. This situation prompted
Russell Stevens (6) to conclude that the simplicity and
general effectiveness of the host resistance and chemical
control approaches had drawn attention away from cultural
control to the point that it enjoyed less popular understanding and support. The plant pest control literature of
the 1900-1965 period demonstrates clearly a preoccupation
with the development of better chemicals and better resistant
varieties (principally against pathogens). Comparatively
little attention was given to pathogen or insect ecology and
cultural controls.

EARLY ADVOCATES OF AN ECOLOGICAL APPROACH TO PEST CONTROL

As the agricultural experiment stations emerged in the
United States in the second half of the nineteenth century,
entomologists and plant pathologists began to discover
biological explanations for the earlier empirically developed
pest control methodology, which had been restricted largely
to natural and cultural measures, sometimes augmented by
minimal use of the earliest insecticides or fungicides.
Partly by intuitive insight and partly because there was
little choice, leading entomologists advocated an ecological
approach to pest control. In the 1880s Stephen A. Forbes,
State Entomologist of Illinois and Professor of Zoology and
Entomology at the University of Illinois, adopted the word
"ecology" and insisted upon the broad application of
ecological studies in dealing with insect problems of
agricultural crops (7). A number of others concerned with
crop protection also advocated this fundamental approach.

In spite of this early position by leading entomologists,
over the next half-century there was a gradual erosion of the

understanding of the importance of ecology in controlling
insect pests. There were, of course, exceptions to this,
and from time to time a plea was made for the ecological
approach. Charles W. Woodworth, Professor of Entomology at
the University of California, advocated an ecologically based
pest management approach throughout his long career (8).
There were other early advocates of an "ecological approach"
to insect pest control. In 1926 Charles Townsend, influenced
by his experience in Peru, stated that "environmental
investigations furnish the only sure basis for work leading
to the speedy discovery of proper measures for the control
of insects, whether for the suppression of injurious forms
or for the extension of beneficial ones." In 1945, before
the impact of DDT and the organic pesticides, A.E. Michelbacher (9) also stressed the importance of ecology in
insect control.

There was apparently no strong parallel concern among
plant pathologists about the application of ecological
principles to the "management" of plant diseases. After
"cause and effect" relationships between pathogens and
disease symptoms were established, it was generally recognized
that pathogen life cycles should be understood to reveal a
"weak link" that might be exploited for control. But
probably because plant pathologists were trained as botanists,
they were not as concerned about the interaction of pathogen
populations and their total environment as were some
entomologists on behalf of insect populations.

EARLY PEST MANAGEMENT FOR THE COTTON BOLL WEEVIL

An analysis of the development of insect control in
cotton also reveals the early foundations of integrated pest
management. The boll weevil, a native of Mexico, entered the
United States in the late 1890s. Gradually the pest spread
from its point of entry in south Texas into other states, and
by 1922 it was distributed almost throughout the entire Cotton
Belt, from Texas to the east coast (10). Research designed to
control or eradicate the pest was begun in 1891 by an
entomologist at the A&M College of Texas (now Texas A&M
University) and in 1894 by USDA entomologists (10). The
early workers recognized the boll weevil problem as an
extremely complex and serious one. Regardless of their
initial notions about dealing with the pest, they soon
rejected eradication as a realistic goal and commenced to
develop what gradually evolved into a highly sophisticated
system of pest management. It is difficult to determine
whose influence was most significant in developing this

system. L.O. Howard (11), incorporating the results of field
studies by Schwarz and Townsend, undoubtedly influenced
others to look at multicomponent suppression techniques. He
stressed cultural control, especially the early fall
destruction of cotton plants, and recommended early planting
and clean cultivation. He also encouraged trapping weevils
late in the fall and overwintering weevils early in the
spring, and destruction of volunteer plants. Malley (12),
Hunter (13,14), Hunter and Hinds (15), and several others
(16) further expanded the multicomponent management approach.

   Although the term "pest management" was not mentioned
in any reference on boll weevil prior to about a decade ago,
the early workers not only preached but also practiced this
approach. By 1901, a systems approach to cotton insect
management had been advocated, and many components of the
weevil's life system and the cotton agroecosystem as well
were understood. By 1904, Hunter and Hinds had presented a
fairly good conceptual model of the pest's life system and
recognized many of the interdependent environmental factors
causing a seasonal change in population density. A highly
complex and sophisticated system had been fully developed
as early as 1920, at which time economic thresholds were
determined as guidelines for beginning treatments with
calcium arsenate. Many of these early management strategies
were highly efficient and were adopted by the more progressive cotton farmers of the day.

SHIFT TO DEPENDENCE ON CHEMICALS

   In spite of the occasional warnings about the hazards of
unilateral approaches to pest control, crop protection in the
United States since 1920 has gradually shifted toward
dependence on chemical pesticides and disease-resistant
varieties. The signals from populations of pests resistant
to chemicals (e.g., red scale resistant to HCN on citrus and
codling moth to lead arsenate on apples) were ignored. The
pattern of developments with cotton insect control during
the period 1920-1945 is a good example of what happened (3).

   It is difficult to pinpoint the causes of the paralysis
of applied cotton entomology that began in the 1920s and
peaked just a few years ago. Applied entomology as a whole
suffered the same fate at about the same time. Perhaps the
cause was a social one, and the remedy lay in public policies
that were beyond the grasp of the age. On the other hand,
the attitudes and mistakes of the entomologists themselves
contributed more than any other single factor.

   The beginning coincided with the time that a number of

cotton entomologists discovered practical methods for applying calcium arsenate. These entomologists dropped their ecologically based work on cultural control, biological control, and resistant varieties, and began exhaustive research on dusting schedules, dosages, swath patterns, and nozzle orifices. This shift in research emphasis paid tremendous dividends, because control of cotton insects with chemicals was spectacular. The applied entomologist became obsessed by the immense power he commanded over nature with these potent chemical weapons. The early 1920s to middle 1940s were definitely dominated by applied entomologists who adhered to and practiced this chemical approach. There were a few dissenters [notably, (17,18)], who warned that the inorganic insecticides should be applied only as necessary to supplement other controls. Nevertheless, the unilateral insecticide approach dominated.

IMPACT OF ORGANIC PESTICIDES

The prevailing philosophy adhered to by applied entomologists during the second quarter of this century was given even greater opportunities for expression when the post-World War II organic insecticides were introduced in the late 1940s. Entomologists enthusiastically adopted into their control programs DDT and other organochlorines and later organophosphorus and carbamate materials.

Whitcomb (19) and Newsom (20) have described the approach that a majority of the cotton entomologists in the post-War period advocated for the new organic materials. In general, entomologists recommended that farmers spray their cotton once weekly from the time it started squaring until near harvest, but there was no method suggested by which the farmer could determine if the spray was actually needed. Certainly, some entomologists had more influence than others in promoting these "overkill" procedures for control of cotton insects. Nevertheless, this approach was favorably accepted by most entomologists of the day. Articles by Rainwater (21), Gaines (10,22), Curl and White (23), and Ewing (24) expressed the general philosophy that prevailed during the first 5 to 15 years following introduction of the organochlorines.

Despite the prevailing insect control philosophy, there also were some very sound insect management programs developed during this period. A classical example was the cultural control program developed for pink bollworm, *Pectinophora gossypiella* Saunders, recently reviewed by Adkisson (25). Also, several entomologists continued to

encourage farmers or scouts to check the cotton fields
regularly and to apply insecticides only when the pest
population reached economic thresholds (26-28). Newsom
and Smith (29) and a few of their followers gathered evidence
that the new organic insecticides seriously affected the
populations of several natural enemies. In general, however,
the period from the late 1940s to the early mid-1960s marked
a time when most major cotton growing areas troubled with
severe insect pests came under a heavy blanket of insecticides.

The rest of the story has been documented recently in
publications by numerous authors (20,25,30-39). All of
these articles pointed to the problems that eventually arose
as a result of overuse of the insecticides introduced after
World War II.

RETURN TO ECOLOGICAL APPROACHES

The pest problems of agricultural crops in the United
States have been aggravated and intensified by a complex of
factors. Some of the factors were related to the limited
base of tactics employed (primarily chemicals and host
resistance). In many instances these no longer controlled
the target pests, but interfered with the control of other
pests and released some species from existing natural control
so that they became pests. In some cases, the chemicals
modified the physiology of the crop plants unfavorably,
created hazards to human health, destroyed pollinators
and other desirable wildlife, and in other ways produced
undesirable effects.

The introduction of new pest species into the U.S.
agroecosystems also placed stress on an already overburdened
pest control technology. Furthermore, the pressure on
agroecosystems toward greater intensity of production over
the years forced them to evolve very rapidly and created new
environments for the pests. As a result, the agroecosystems
often became more vulnerable to pests. Changes in tillage,
water management, crop varieties, fertilization, and other
agronomic practices greatly influenced pest incidence, very
often in favor of the abundance of the pest species. The
increased complexity and intensity of agricultural production
practices along with reduced genetic diversity in many
agricultural crop species combined to produce a new
magnitude of crop hazards from pests.

The intensification and increasing complexity of crop
protection problems, coupled with the associated

environmental, financial, and health hazards of heavy chemical
usage combined to stimulate great interest in the importance
of crop protection and in the broad ecological approach as a
sound approach to acceptable solutions. Also of great
importance was the increased financial support for pest
management research, extension, and field-implementation
programs.

As the problems intensified in crop protection, the
debate over the matter also intensified. This finally
erupted in the *Silent Spring* episode (40,41). The President's
Science Advisory Committee issued a special report in 1963
entitled *Use of Pesticides* that found fault with a number of
crop protection chemicals, especially insecticides. The
Southern Corn Leaf Blight epidemic of 1970 (42) emphasized the
problem of the genetic vulnerability of major crop species
in the United States to damaging attacks by pests (43).
These situations, combined with increased awareness of a
world food crisis, motivated government and institutional
actions supportive of the development of integrated pest
management systems for major agroecosystems in the United
States.

A major step toward development of integrated pest
management programs was taken by the federal government in
1972. In his Message on Environmental Protection, the
President of the United States directed the cognizant agencies
of government to take immediate action toward development of
pest management programs in order to protect: (1) the nation's
food supply against the ravages of pests, (2) the health of
the population, and (3) the environment. The President's
Directive prompted funding of a national research project
involving 19 universities and various federal agencies
entitled "The Principles, Strategies, and Tactics of Pest
Population Regulation and Control in Major Crop Ecosystems."
This NSF/EPA Integrated Pest Management Project also known
as the "Huffaker Project" has taken the lead role in providing
the mechanisms, i.e., "the way-of-doing business," in the
complex milieu of multidisciplinary plant protection as a
component of crop production. The "Huffaker Project" has
shown the way toward a new era in plant protection.

The speakers in the remainder of this Symposium will
detail the many advances in integrated pest management that
have come from this project in the past five years.

Other programs initiated in 1972 were pilot projects for
implementing pest management programs in the various states;
curriculum development for training and certification of crop
protection specialists by the land-grant universities; and
pilot pest management research projects within the USDA's

Agricultural Research Service in collaboration with state groups. These actions were paralleled with an intensification of pest management research within state agricultural experiment stations and federal agencies financed by both state and federal sources.

The foregoing discussion has traced the history of integrated pest management back into the nineteenth century and even earlier. The concept of integrated pest management is not a disjunct development in crop protection, but is an evolutionary stage in the history of pest control. It is significantly different from earlier strategies in that it represents a new conceptual approach that sets crop protection in a new context within a crop production system (44). Many components of the integrated pest management concept were developed in the late nineteenth and early twentieth centuries, but integrated pest management as it is now conceived is unique because it is based on ecological principles and integrates multidisciplinary methodologies in developing agroecosystem management strategies that are practical, effective, economical, and protective of both public health and the environment. The efforts of early crop protection specialists to control pests with ecologically based cultural methods were not satisfactory; consequently, entomologists, plant pathologists, and later, weed scientists, became preoccupied with the discovery and development of pesticides that were economical and effective. Unfortunately, chemical methods were often not used to supplement cultural and biological methods of control but to supplant them. Our state of technology and understanding of host-pest interactions has now evolved to the point that an integration of pest control tactics for the control of a given class of pest (e.g., insects, plant pathogens, etc.) and for multiple classes of pests is not only feasible but necessary, given the inadequacies of single-method, single-discipline approaches and their potential for undesirable effects on nontarget beneficial and pest species.

REFERENCES

1. Smith, R.F. and W.W. Allen. 1954. Insect control and the balance of nature. Sci. Am. 190(6):38-42.
2. Stern, V.M., R.F. Smith, R. van den Bosch, K.S. Hagen. 1959. The integrated control concept. Hilgardia 29:81-101.
3. Smith, R.F., J.L. Apple, D.G. Bottrell. 1976. The origins of integrated pest management concepts for agricultural crops. pp. 1-16 *in* Integrated Pest Management. J.L. Apple and R.F. Smith, eds. Plenum, New York.

4. Parris, G.K. 1968. Chronology of Plant Pathology. Johnson and Sons, Starkville, Miss.
5. Lodeman, E.G. 1903. The Spraying of Plants. MacMillan, New York.
6. Stevens, R.B. 1960. Cultural practices in disease control. pp. 357-429 *in* Plant Pathology: An Advanced Treatise. J. G. Horsfall and A.E. Dimond, eds., Vol. 3. Academic Press, New York.
7. Metcalf, C.L. 1930. Obituary, Stephen Alfred Forbes, May 29, 1844-March 13, 1930. Entomol. News 41(5):175-78.
8. Smith, R.F. 1975. The origin of integrated control in California—an account of the contributions of C.W. Woodworth. Pan Pac. Entomol. 50(4):426-29.
9. Michelbacher, A.E. 1945. The importance of ecology in insect control. J. Econ. Entomol. 38:129-30.
10. Gaines, R.C. 1957. Cotton insects and their control. Annu. Rev. Entomol. 2:319-38.
11. Howard, L.O. 1896. The Mexican cotton boll weevil. U.S. Dep. Agr. Bur. Entomol., Circ. No. 14.
12. Malley, F.W. 1901. The Mexican cotton-boll weevil. USDA Farmers' Bull. 130.
13. Hunter, W.D. 1902. The present status of the Mexican cotton-boll weevil in the United States. pp. 369-80 *in* Yearbook of the United States Department of Agriculture-1901. U.S. Govt. Print. Off., Washington, D.C.
14. ————. 1903. Methods of controlling the boll weevil. USDA Farmers' Bull. No. 163.
15. Hunter, W.D. and W.E. Hinds. 1904. The Mexican cotton boll weevil. USDA, Div. Entomol. Bull. No. 45.
16. Dunn, H.A. 1964. Cotton Boll Weevil (*Anthonomus grandis* Boh.): Abstracts of Research Publications, 1843-1960. U.S. Dep. Agr. Coop. State Res. Serv., Misc. Publ. 985.
17. Isley, D. and W.J. Baerg. 1924. The boll weevil problem in Arkansas. Ark. Agr. Exp. Sta. Bull. No. 190.
18. Baerg, W.J., D. Isley, M.W. Sanderson. 1938. Fiftieth Annual Report. Entomology. Ark. Agr. Exp. Sta. Bull. No. 368, pp. 62-66.
19. Whitcomb, W.H. 1970. History of integrated control as practiced in the cotton fields of the south central United States. pp. 147-55 *in* Proc. Tall Timbers Conf. Ecol. Anim. Control Habitat Manage., No. 2.
20. Newsom, L.D. 1970. The end of an era and future prospects for insect control. pp. 117-36 *in* Proc. Tall Timbers Conf. Ecol. Anim. Control Habitat Manage., No. 2.
21. Rainwater, C.F. 1952. Progress in research on cotton insects. pp. 497-500 *in* Insects. The Yearbook of Agriculture 1952. U.S. Govt. Print. Off., Washington, D.C.
22. Gaines, R.C. 1952. The boll weevil. pp. 501-04 *in* Insects.

The Yearbook of Agriculture 1952. U.S. Govt. Print. Off.,
Washington, D.C.

23. Curl, L.F. and R.W. White. 1952. The pink bollworm. pp.
    505-11 *in* Insects. The Yearbook of Agriculture 1952. U.S.
    Govt. Print. Off., Washington, D.C.

24. Ewing, K.P. 1952. The bollworm. pp. 511-14 *in* Insects.
    The Yearbook of Agriculture. 1952. U.S. Govt. Print. Off.,
    Washington, D.C.

25. Adkisson, P.L. 1972. Use of cultural practices in insect
    pest management. pp. 37-50 *in* Implementing Practical Pest
    Management Strategies. Proceedings of the National Exten-
    sion Insect-Pest Management Workshop, Purdue University,
    March 14-16.

26. Smith, G.L. 1953. Supervised control of cotton insects.
    in Conference on Supervised Control of Insects, Depart-
    ment of Entomology and Parasitology, University of
    California, Berkeley, February 6-7.

27. Smith, R.F. 1949. Manual of supervised control. Division
    of Entomology and Parasitology, University of California,
    Berkeley.

28. Boyer, W.P., C. Lincoln, L.O. Warren. 1962. Cotton
    scouting in Arkansas. Ark. Agr. Exp. Sta. Bull. No. 656.

29. Newsom, L.D. and C.E. Smith. 1949. Destruction of certain
    insect predators by applications of insecticides to
    control cotton pests. J. Econ. Entomol. 42:904-08.

30. Adkisson, P.L. 1969. How insects damage crops. Conn. Agr.
    Exp. Sta. Bull. 708:155-64.

31. ———. 1971. Objective uses of insecticides in agricul-
    ture. pp. 110-20 *in* Proceedings of Symposium on Agricul-
    tural Chemistry-Harmony or Discord for Food, People and
    the Environment. J.E. Swift, ed. University of California,
    Division of Agricultural Science.

32. ———. 1973. The integrated control of the insect pests
    of cotton. pp. 175-88 *in* Proc. Tall Timbers Conf. Ecol.
    Anim. Control Habitat Manage., No. 4.

33. ———. 1973. The principles, strategies and tactics of
    pest control in cotton. pp. 274-83 *in* Insects: Studies
    in Population Management. P.W. Geier, L.R. Clark, D.J.
    Anderson, H.A. Nix, eds. Ecol. Soc. Aust. (Mem. 1)
    Canberra.

34. Smith, R.F. and R. van den Bosch. 1967. Integrated con-
    trol. pp. 295-340 *in* Pest Control: Biological, Physical,
    and Selected Chemical Methods. W.W. Kilgore and R.L.
    Doutt, eds. Academic Press, New York.

35. Smith, R.F. 1969. The new and the old in pest control.
    Proc. Accad. Nazion. dei Lincei, Rome (1968) 366(138):
    21-30.

36. ———. 1970. Pesticides: their use and limitations in

pest management. pp. 103-11 *in* Concepts of Pest Management. R.L. Rabb and F.E. Guthrie, eds. North Carolina State University, Raleigh.

37. ————. 1971. Economics of pest control. pp. 53-83 *in* Proc. Tall Timbers Conf. Ecol. Anim. Control Habitat Manage., No. 3.

38. Doutt, R.L. and R.F. Smith. 1971. The pesticide syndrome. pp. 3-15 *in* Biological Control. C.B. Huffaker, ed. Plenum, New York.

39. Stern, V.M. 1969. Interplanting alfalfa in cotton to control lygus bugs and other insect pests. pp. 55-69 *in* Proc. Tall Timbers Conf. Ecol. Anim. Control Habitat Manage., No. 1.

40. Carson, R. 1962. Silent Spring. Hamish Hamilton, London.

41. Wescott, C. 1962. Halftruths or whole story, a review of Silent Spring. Manufacturing Chemist's Association, Washington, D.C.

42. Tatum, L.A. 1971. The southern leaf blight epidemic. Science 171:1113-16.

43. National Academy of Sciences. 1972. Genetic Vulnerability of Major Crops. Committee on Genetic Vulnerability of Major Crops, NAS, Washington, D.C.

44. Brader, L. 1976. Plant protection and production in modern society. EPPO Bull. 6(4):249-63.

DISCUSSION

UNIDENTIFIED: Your definition of "insect pest management" and "insect pest control" is fine, but the terms have been expropriated by the chemical industry, with the so-called "dusties" calling themselves "insect pest managers." In the cotton project, for example, this terminology was used to sell chemicals to kill lygus bugs, which van den Bosch has called a "ghost pest."

R.F. SMITH: We don't always agree with van den Bosch. Also we don't agree that everyone who calls himself an insect pest manager actually practices IPM. All we can hope for is to educate these people; we can't control the use of the term. There are claims coming from all directions.

UNIDENTIFIED: Could you comment on the possible impact of the Gregorio bill in California, which would prevent pesticide salesmen from advising farmers?

R.F. SMITH: Conflicts of interest make it difficult for pesticide salesmen to give unbiased advice. There have been many attempts to limit this, but legislation is not enough; we need to educate farmers, extension agents, and others.

SOCIOECONOMIC AND LEGAL ASPECTS OF PEST CONTROL

David Pimentel

Department of Entomology and
Section of Ecology and Systematics
Cornell University
Ithaca, New York

INTRODUCTION

The world population today faces food shortages. It is estimated that from 1/2 to 1 billion humans are protein/calorie malnourished (1). The food supply situation is not expected to improve, because a rapidly growing human population will need much larger quantities of food.

Many constraints face us in devising ways to solve the food problem. One, of course, is the amount of arable land available. Because most of the arable land is already in production, our challenge is to obtain greatly increased yield from about the same amount of cropland (2,3). To accomplish this will add a serious stress on our supply of fossil energy.

One important strategy that should be included in the management system is the substantial reduction in losses of food crops to pests. On a worldwide basis, crop losses to all pests, including insects, diseases, and weeds, are estimated to be nearly 48%. This includes 35% preharvest losses and 20% postharvest losses (4,5). Crop losses to pests in the United States are only slightly less, at 33% preharvest losses and 9% postharvest losses (5,6).

Because of these substantial losses of valuable food and feed crops in the United States and the world, we need to assess present methods of pest control as an integral part of the total system (fig. 1). In this paper I will focus on the socioeconomic and legal aspects of pest control, although they should be recognized as only one part of any systems management program for pest control.

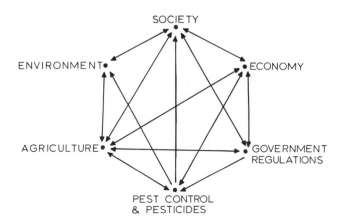

SOCIETY

ENVIRONMENT

ECONOMY

AGRICULTURE

GOVERNMENT
REGULATIONS

PEST CONTROL
& PESTICIDES

Fig. 1. The interdependencies of pest control as a part of the total system.

SOCIOECONOMIC ASPECTS OF PESTS AND PEST CONTROL

Crop Losses to Pests

Currently, an estimated 33% of all crops in the United States is lost to pests (13% to insects, 12% to pathogens, and 8% to weeds), in spite of all pesticide and bioenvironmental controls used (5,6).

According to survey data collected from 1942 to the present, losses from weeds have declined, probably because of improved mechanical cultivation and herbicide weed control technologies. In that same period, however, the losses from plant pathogens (including nematodes) have increased from 10.5% to about 12% (Table 1).

Crop losses from insect pests increased nearly 2-fold (7% to about 13%) from the 1940s to the present. This increase occurred in spite of a 10-fold increase in insecticide use. The impact of these losses has been lessened by increased crop yields through the use of higher-yielding crop varieties and increased energy inputs in the form of fertilizers (7).

This alarming increase in crop losses due to insect damage, despite increased insecticide use, can be accounted for by some of the changes that have taken place in agricultural production. These include: (1) the planting of crop varieties that are susceptible to insect pests; (2) reduced crop rotations and crop diversity with increased reliance on continuous culture of the same crop (e.g., corn) (8,9);

Table 1. Comparison of Annual Pest Losses in Agriculture for the Periods 1904, 1910-35, 1942-51, and 1951-60 and an Estimate of Losses for 1974 [after (5)].

	Insects		Diseases		Weeds		Total Loss		Potential Production
	$	%	$	%	$	%	$	%	$
1974	7.2	13	6.6	12	4.4	8	18.2	33	55
1951-60	3.8	12.9	3.6	12.2	2.5	8.5	9.9	33.6	29.5
1942-51	1.9	7.1	2.8	10.5	3.7	13.8	8.4	31.4	26.7
1910-35	0.6	10.5	NA[a]	NA	NA	NA	NA	NA	5.7
1904	0.4	9.8	NA	NA	NA	NA	NA	NA	4.1

[a]Not available

(3) reduced sanitation, including destruction of infected fruit and crop residues (e.g., apples); (4) reduced tillage, with more crop remains left on the land surface (e.g., corn) (10); (5) culturing crops in climatic regions where they are more susceptible to insect attack (e.g., potatoes, broccoli) (11); (6) increased pesticide resistance in insects (12-15); (7) destruction of natural enemies of certain pests resulting in the need for additional pesticide treatments (e.g., cotton) (16); (8) use of pesticides that alter the physiology of some crop plants, making them more susceptible to insect attack (e.g., corn) (17,18); and (9) reduced FDA tolerance and increased "cosmetic standards" by processors and retailers for fruits and vegetables (19).

Pest Control Technologies

Although much emphasis is given pesticide usage in U.S. agriculture, the dominant pest control technologies in the United States are based on "bioenvironmental" (nonchemical) controls. In fact, bioenvironmental controls are employed on a larger percentage of agricultural acres than are pesticide controls (5,20). "Bioenvironmental" control is defined as any nonchemical control method utilized to reduce pest populations by environmental manipulations and biological control (20).

For insect control, bioenvironmental controls are employed on about 9% of the crop acres (5), whereas insecticidal controls are used on only 6% of U.S. crop acreage (21). For the control of plant diseases, some form of bioenvironmental

control, primarily host plant resistance, is employed on
about 95% of the acreage; fungicide treatments are used on 1%
of the acres (21). For weed control, bioenvironmental controls, primarily mechanical cultivation, are used on an
estimated 80% of the crop acreage (5,22); herbicides are used
on 17% of the acres (21).

Although bioenvironmental controls are used much more
frequently than are pesticide controls, more than 800 million
pounds of pesticides are used annually (23) (fig. 2). Of this
total, an estimated 550 million pounds of pesticide are used
in agricultural production, and the remainder is applied by
homeowners, industry, and government agencies.

Pesticide use in agriculture is not evenly distributed
(Table 2). For example, 50% of all insecticide used in
agriculture is applied to the nonfood crops of cotton and
tobacco. Of the food crops, corn, fruit, and vegetables
receive the largest amounts of insecticide. Of the herbicidal
material applied, 45% is used on corn, with the remaining 55%
distributed among numerous other crops (Table 2). Most of the
fungicidal material is applied on fruit and vegetables, with
only a small amount used on field crops (Table 2).

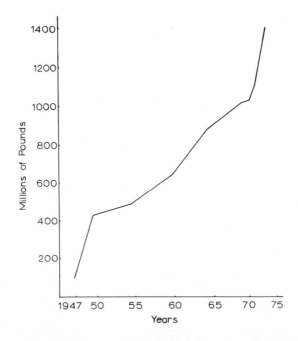

*Fig. 2. Estimated amount of pesticide produced in the
United States (24,25).*

Table 2. Some Examples of Percentages of Crop Acres Treated, of Pesticide Amounts Used on Crops, and of Acres Planted to this Crop (21,26).

Crops	Insecticides		Herbicides		Fungicides		% of Total Crop Acres
	% Acres	% Amount	% Acres	% Amount	% Acres	% Amount	
Nonfood	NA[a]	50	NA	NA	< 0.5	NA	NA
Cotton	61	47	82	9	4	1	1.11
Tobacco	77	3	7	NA	7	NA	0.11
Field Crops	NA	33	NA	NA	NA	15	NA
Corn	35	17	79	45	1	NA	7.43
Peanuts	87	4	92	2	85	11	0.16
Rice	35	1	95	3	0	NA	0.22
Wheat	7	1	41	5	0.2	NA	6.11
Soybeans	8	4	68	16	2	NA	4.19
Pasture Hay & Range	0.5	2	1	4	0	NA	68.40
Vegetables	NA	7	NA	2	NA	24	NA
Potatoes	77	2	51	NA	49	10	0.16
Fruit	69	9	NA	1	NA	60	NA
Apples	91	3	35	NA	61	18	0.07
Citrus	72	2	22	NA	47	24	0.08
All Crops	6	35	17	51	0.9	14	100

[a]Not available

59

Benefits of Pesticide Controls

In spite of the estimated 800 million pounds of pesticides and bioenvironmental controls used in agriculture, an
estimated 33% of all crops are lost annually due to pest
attack in the United States (fig. 2). This loss of food and
fiber amounts to about $35 billion, or enough to pay for our
1976 oil imports.

One question frequently asked is, what would our crop
losses to pests be if all pesticides were withdrawn from use,
and readily available nonchemical control methods were substituted where possible? We are in the process of examining
this question, and it appears that crop losses based on dollar
value would increase from the estimated 33% to about 42% (27).
Thus, we estimate that a 9% increase in crop losses could be
anticipated if pesticides were withdrawn from use.

These estimated dollar losses increase somewhat when all
nonfood crops such as cotton, tobacco, hay, and pasture are
excluded from the analysis, leaving only food crops. Based
only on food crops, the loss of crops grown without pesticides
is estimated to increase from 9% to 11% (27).

Crop losses without pesticides were also evaluated based
on food and feed energy, expressed as kilocalories (kcal)(27).
Crop losses from insects, diseases, and weeds grown without
pesticides but using some alternative controls were estimated
to increase only 1% (27). When nonfood crops were excluded,
the increased loss of crops grown without pesticides was only
4% in food calories (27).

Based on these estimates of increased food energy losses,
1% for total crops or 4% for food crops, there would be no
serious food shortages in the United States if crops were not
treated with pesticides. Although the supply of food in the
nation would be ample, the quantities of certain fruits and
vegetables, such as apples, peaches, plums, onions, tomatoes,
peanuts, and certain other crops would be significantly reduced (27). Because of this, some fruits and vegetables that
we are accustomed to eating might have to be replaced with
others.

Although our food-energy supply would be little affected
by the withdrawal of pesticide use, the dollar loss to the
nation would be considerable. This would amount to an
estimated $8.7 billion loss, including added costs of employ-
ing alternative nonchemical controls that would be used if
pesticide use were withdrawn (27). Considering that current
pesticide treatment costs, material and application, are
estimated to be $2.2 billion, the return per dollar invested
in pesticide control is about $4 (27).

Impact of Pesticides on the Environment and Public Health

In calculating the benefits of pesticides at $4 per dollar invested in control, Pimentel et al. (27) did not include a dollar value for the "external costs" of human poisonings and the impact of pesticides on the environment. To evaluate the external costs of pesticide use, the relationship we have with our environment must be understood.

Although everyone knows why food is essential, not everyone is aware of why the environment is equally essential to us. We *cannot* maintain our high standards of health and achieve a quality life in an environment consisting only of our crop plants and livestock. Most of the estimated 200,000 species of plants and animals in the United States are an integral and functioning part of our ecosystem. Many of these species help renew atmospheric oxygen. Some prevent us from being buried by human and agricultural wastes and others help purify our water. Trees and other vegetation help maintain desirable climate patterns. In addition some insects are essential in pollinating forage, fruit, and vegetable crops for high yields. No one knows *how much* the population numbers of these 200,000 species could be reduced and/or *how many* species could be eliminated before agricultural production and public health would be threatened.

The impact of pesticides on agriculture itself, the environment, and public health is significant. For example, the Environmental Protection Agency (28) reports that "as many as 14,000 individuals may be non-fatally poisoned by pesticides in a given year" and 6,000 are injured seriously enough to require hospitalization. Associated with the poisonings are an estimated 200 fatalities annually. Apparently none of the poisonings and fatalities were due to eating food crops that were treated properly with pesticides. The people especially prone to pesticide poisoning are pesticide production workers, farm field workers and pesticide applicators (29). Of the field workers and pesticide applicators in the United States, an estimated 2,826 were hospitalized in 1973 because of pesticide poisoning (30).

Because of their widespread use, pesticides are consumed by people. In fact, in one study from 93% to 100% of the people surveyed tested positive for one or more pesticides (31). Annual studies conducted by the FDA determine the kinds and amount of insecticide residues in typical human daily diets (32,33). The residues of DDT and its metabolites in foods generally have been low (95% below 0.51 ppm). The incidence of contamination, however, was high, i.e., about 50% of the food samples contained minute but detectable insecticide residues (33). Residues of the phosphate and carbamate

insecticides occurred less frequently and at relatively lower
dosages because these insecticides are generally less persistent than are the chlorinated insecticides (32-34). The FDA
data suggest, however, that residues of phosphate and carbamate insecticides are beginning to increase in raw products,
and therefore also in the total diet. This increase was to
be expected after DDT was banned by the EPA in 1972. In
addition, the public will continue to be exposed to residues
of DDT and other chlorinated pesticides because these persist
in the environment.

At present, overall pesticide residue levels appear to
be sufficiently low to present little or no danger to human
health in the short term (35). Samples of fruits and vegetables rarely have insecticide residues that exceed 2 ppm (33,
34). For example, of 1,551 samples of "large fruit" only 10
showed residues of 2.8 to 13.5 ppm. Residues ranging from
2.3 to 84.0 ppm were detected in 97 out of 2,461 leafy and
stem vegetable samples (33). Unfortunately, little is known
about the effects long-term, low-level dosages of pesticides
may have on public health (35). Furthermore, the possible
interaction between low-level dosages of pesticides and the
numerous drugs and food additives the public consumes has not
been completely studied.

The ecological effects of pesticides on nontarget species
are varied and complex. For example, some pesticides have
influenced the structure and function of ecosystems, reduced
species population numbers in certain regions, or altered the
natural habitat under some conditions. Some have changed the
normal behavioral patterns in animals, stimulated or sup-
pressed growth in animals and plants, or modified the reproductive capacity of animals. In addition some have altered
the nutritional content of foods, increased the susceptibility
of certain plants and animals to diseases and predators, or
changed the natural evolution of species populations in some
regions (15). Because of this great variation in effect, it
is necessary to study the impact of individual pesticides to
obtain a fair and balanced picture.

Another interesting aspect of the pesticide problem is
the fact that the 800 million pounds of pesticide applied
in the United States are used to control only about 2,000 pest
species. If these pesticides reached only the target species,
the pollution problem would not be of concern. Unfortunately,
however, only about 1% of the pesticide used ever hits the
target pests (20). Often as little as 25% to 50% of the pesticide formulation reaches the crop area, especially when pesticides are applied by aircraft (36-39). Considering that
about 65% of all agricultural insecticides are applied by
aircraft (40), the risk both to the environment and to public

health is great.

In summary, the main reasons for serious ecological problems with pesticides are: (1) pesticides are biological poisons (toxicants); (2) large quantities are applied to the ecosystem annually; and (3) application technology is inefficient. Much scientific evidence substantiates the concern that pesticides have caused measurable damage to many species of birds, fishes, and beneficial insects (14,41).

We estimate that the external cost of pesticide use is at least $3 billion annually in the United States. Included in this external cost estimate are: hospitalization costs for 6,000 human pesticide poisonings (28); costs of about 60,000 days of work lost because of these pesticide poisoning hospitalizations; additional medical costs for 8,000 human pesticide poisonings treated as outpatients; and costs of about 30,000 days of work lost from humans not ill enough to be hospitalized. (Note, consideration of money costs of the estimated 200 human deaths (28) were not included in the calculation.) Other "external costs" included were: about $12 million in direct honey bee losses[1]; reduced fruit crops and reduced pollination from the destruction of wild bees and honey bees; livestock losses; commercial and sports fish losses; bird and mammal losses; natural enemies of pests destroyed, resulting in outbreaks of other pests; pest problems that result from pesticide effects on the physiology of crop plants; and increased pesticide resistance in pest populations. All of these contribute to the external costs of pesticides and must be considered in any cost/benefit analysis.

LEGAL ASPECTS OF PEST CONTROL

Pesticide use and pest controls are intimately involved in and affected by state and federal legislation. A few selected examples of legislation and governmental actions will illustrate the complexity of the legal aspects of pest control.

Government Price Supports and Acreage Controls

Government price supports and acreage controls were implemented for cotton and other crops to stabilize agricultural production and to protect the farmer (42). Although

---

[1]R.A. Morse, *personal communication*

they have accomplished some of the desired aims, government
price supports and acreage controls often have resulted in
increasing pesticide use and production costs (9,42,43).

For example, government acreage controls for cotton tend
to encourage some cotton production in regions where production costs are relatively high and where much insecticide
needs to be used for the control of pests (9,43). Without
government acreage controls, generally cotton could be grown
in those regions best suited for its production. National
cotton production would be more efficient, and production
costs were estimated at 36% less without controls than when
cotton was produced under acreage controls (42,43).

In addition, producing cotton without acreage controls
and in the best suited regions reduced the estimated use of
insecticides by about 40% (9). Cotton production without
acreage controls would shift away from the southeast and
concentrate more in the western and southwestern regions (9).
The southeastern region is the region where the cotton boll
weevil is a most serious pest. Treating for the boll weevil
results in outbreaks of the cotton bollworm and budworm,
because their natural enemies are destroyed by insecticides
(44,45).

Pesticide Residues

Before 1900, the aim of pest control was to exterminate
the pest by any means short of destroying the crop. For
example, lead arsenate was used in large quantities for insect
control, and it was common to observe fruits and vegetables
for sale that were "powder white" with residues.

Concern for the health of humans consuming these contam-
inated foods developed during the early 1900s and various
government regulations slowly emerged. In 1954, tolerances
were established for pesticides on raw agricultural commodities and were regulated by the Federal Food, Drug and Cosmetic
Act as amended by the Miller Bill of 1954. This legislation
limited the quantity of pesticide residues found in or on
fruits, vegetables, and other agricultural products.

Therefore, by the mid-fifties human health became a
significant factor in assessing the risks and benefits of
pest control recommendations. Since then there has been
increasing concern for human health, because medical research
has confirmed that in addition to direct toxicity some pesticides are carcinogenic. As a result, the standards in recent
pesticide registration procedures are more stringent than
those in the initial legislation.

In addition, public concern, stirred in part by Rachel

Carson's (46) book, *Silent Spring,* focused on the deterioration of the environment as well as public health problems caused by pesticides. The public has demanded, and rightly so, to see the pesticide "balance sheet" on costs and benefits. This demand brought new pesticide legislation. In 1970 the Environmental Protection Agency (EPA) was established to enforce protection of the environment from all pollutants, including pesticides. All previously registered pesticides are now being carefully reevaluated by EPA for registration under the 1972 amendments to the Federal Insecticide, Fungicide and Rodenticide Act. The task is tremendous.

Costs of Pesticide Development

Public concern for pesticides, and the new government legislation, have resulted in more rigorous testing of pesticides before they are cleared for use on crops. One measure of the impact of such regulations is the amount of money now required to develop and register a pesticide for use. In 1945, pesticide manufacturers spent about $1 million to develop and market a pesticide, whereas today they spend between $6-12 million (47). Pesticides are not the only product or activity subjected to increased regulations; indeed, during the last quarter century, most aspects of society have witnessed added regulations. With the rapidly increasing human population, there are larger urban centers, more automobiles, more factories, and more power plants, etc. So, too, there are more pesticides (fig. 2) and other chemicals introduced into our ecosystem than ever before. All these have a definite impact on the quality of our environment, our supply of such finite resources as fuel and land, as well as the health status of our people.

FDA and Insects in Foods

Over the years, the Food and Drug Administration (FDA) has acted to reduce the tolerance levels of insects and insect parts that are permitted in food. This has resulted in an increased use of pesticides in agriculture (19). For instance, the established tolerance levels for spinach leaf miners during the past 40 years have declined from 40 to 9 per 100 grams of spinach and aphids from 110 to 60 (48,49). To attain these standards, 300% more insecticide is used on spinach than in the past (1930) (19). Similar increases in pesticide use have occurred in processed apples, cherries, strawberries, broccoli, tomatoes, and other fruits and vegetables.

FDA has reduced these tolerance levels for small herbivorous insects in foods without ever demonstrating that these small insects create a health hazard. In fact, all past studies and reports document that these small insects contribute to the nutritional value of the vegetables and fruits (19).

Furthermore, wholesalers and retailers have raised "cosmetic standards" thus increasing the percentage of crops that must be discarded because of slight pest damage. Although such fruits and vegetables have a slightly less-than-perfect outer appearance, they are completely safe and nutritionally sound. In an effort to prevent such loss, the farmer often uses more insecticides.

The actions by FDA and wholesalers and retailers have resulted in increasing insecticide usage on fruits and vegetables an estimated 10% to 20% (19). The costs associated with these actions of the FDA and wholesalers and retailers include increasing health hazards from insecticides, reducing environmental quality, using more energy, and increasing food costs to the consumer.

CONCLUSION

Based on current trends, I expect pesticide use, particularly herbicide use, to increase for at least the next decade. At the same time that pesticide use expands, I project that the direct economic benefits of pesticide use by growers and others will decline. The benefits to users will decline as they have to pay for a greater share of the environmental and public health costs associated with pesticide use.

In response to continued public concern about pesticide poisoning and environmental pollution, more legislation and regulations to reduce the hazards associated with growing pesticide use are inevitable.

The real breakthrough in pest control technology, however, will occur when pest control becomes an integral part of total management systems programs for agricultural production. This systems approach will be developed and put into action only when pest control costs and other crop production costs become burdensome, and our desire for conserving environmental resources becomes a serious concern.

REFERENCES

1. National Academy of Sciences. 1977. World Food and Nutrition Study. (Manuscript).

2. Pimentel, D., W. Dritschilo, J. Krummel, J. Kutzman. 1975. Energy and land constraints in food protein production. Science 190:754-61.

3. Pimentel, D., E.C. Terhune, R. Dyson-Hudson, S. Rochereau, R. Samis, E.A. Smith, D. Denman, D. Reifschneider, M. Shepard. 1976. Land degradation: effects on food and energy resources. Science 194:149-55.

4. Cramer, H.H. 1967. Plant protection and world crop production. Pflanzenschutznachrichten 20(1):1-524.

5. Pimentel, D. 1976. World food crisis: energy and pests. Bull. Entomol. Soc. Am. 22:20-26.

6. United States Department of Agriculture. 1965. Losses in Agriculture. Agr. Handb. No. 291, U.S. Govt. Print. Off., Washington, D.C.

7. Pimentel, D., L.E. Hurd, A.C. Bellotti, M.J. Forster, I. N. Oka, O.D. Sholes, R.J. Whitman. 1973. Food production and the energy crisis. Science 182:443-49.

8. Pimentel, D. 1961. Species diversity and insect population outbreaks. Ann. Entomol. Soc. Am. 54:76-86.

9. Pimentel, D., C. Shoemaker, E.L. LaDue, R.B. Rovinsky, N. P. Russell. 1977. Alternatives for reducing insecticides on cotton and corn: economic and environmental impact. Report on Grant No. R802518-02, Environmental Protection Agency, Washington, D.C.

10. Musick, G.J. and H.B. Petty. 1973. Insect control in conservation tillage systems. pp. 120-25 *in* Conservation Tillage. Soil Cons. Soc. Am.

11. Pimentel, D. 1977. Ecological basis of insect pests, pathogen and weed problems. pp. 3-31 *in* Origins of Pest, Parasite, Disease and Weed Problems. J.M. Cherrett and G.R. Sagar, eds. Blackwell Scientific Publications, Oxford.

12. Georghiou, G.P. 1972. The evolution of resistance to pesticides. Annu. Rev. Entomol. 3:122-68.

13. Hance, R.J. 1977. Ecological aspects in the long term use of pesticides. Paper presented at Southeast Asian Workshop on Pesticide Management. Bangkok, Thailand.

14. Pimentel, D. 1971. Ecological Effects of Pesticides on Non-Target Species. U.S. Govt. Print. Off., Washington, D.C.

15. Pimentel, D. and N. Goodman. 1974. Environmental impact of pesticides. pp. 25-52 *in* Survival in Toxic Environments. M.A.Q. Khan and J.P. Bederka, Jr., eds. Academic Press, New York.

16. van den Bosch, R. and P.S. Messenger. 1973. Biological Control. Intext Educational Publishers, New York.

17. Dunham, E.W. and J.C. Clark. 1941. Cotton aphid multiplication following treatment with calcium arsenate. J. Econ. Entomol. 34:587-88.

18. Oka, I.N. and D. Pimentel. 1976. Herbicide (2,4-D) increases insect and pathogen pests on corn. Science 193: 239–40.
19. Pimentel, D. E.C. Terhune, W. Dritschilo, D. Gallahan, N. Kinner, D. Nafus, R. Peterson, N. Zareh, J. Misiti, O. Haber-Schaim. 1977. Pesticides, insects in foods, and cosmetic standards. BioScience 27:178–85.
20. President's Science Advisory Committee. 1965. Restoring the Quality of our Environment. Report of the Environmental Pollution Panel, The White House, Washington, D.C.
21. United States Department of Agriculture. 1975. Farmers' use of pesticides in 1971...extent of crop use. Econ. Res. Serv., Agr. Econ. Rep. No. 268.
22. National Academy of Sciences. 1968. Weed Control. Principles of Plant and Animal Pest Control. Vol. 2. Washington, D.C.
23. United States Department of Agriculture. 1976. Agricultural Statistics 1976. U.S. Govt. Print. Off., Washington, D.C.
24. ———, 1971. The Pesticide Review 1970. Agr. Stab. Conserv. Serv., Washington, D.C.
25. Fowler, D.L. and J.N. Mahan. 1975. The Pesticide Review 1974. U.S. Dept. Agr., Agr. Stab. Conserv. Serv., Washington, D.C.
26. United States Bureau of the Census. 1973. Census of Agriculture, 1969. Vol. 5. Special Reports. Parts 1, 4-6. U.S. Govt. Print. Off., Washington, D.C.
27. Pimentel, D., J. Krummel, D. Gallahan, J. Hough, A. Merrill, I. Schreiner, P. Vittum, F. Koziol, E. Back, D. Yen, S. Fiance. 1977. Benefits and costs of pesticide use in U.S. food production. (Manuscript).
28. Environmental Protection Agency. 1974. Strategy of the Environmental Protection Agency for controlling the adverse effects of pesticides. Office of Pesticide Programs, Office of Hazardous Materials, Washington, D.C.
29. California Department of Health. 1974. Occupational disease in California attributed to pesticides and other agricultural chemicals, 1971-1973. Occupational Health Section and Center for Health Statistics, Berkeley, California.
30. Savage, E.P., T. Keefe, G. Johnson. 1976. The pesticide poisoning rate is low. Agrichem. Age, May:15-17.
31. Environmental Protection Agency. 1976. Human monitoring program. Pest. Monitor. Quart. Rep. #6.
32. Corneliussen, P.E. 1972. Pesticide residues in total diet samples VI. Pestic. Monit. J. 5(4):313-30.
33. Duggan, R.E. and M.B. Duggan. 1973. Pesticide residues in food. pp. 334-64 *in* Environmental Pollution by Pesticides. C.A. Edwards, ed. Plenum, London.

34. Food and Drug Administration. 1975. Compliance program evaluation. Total diet studies: FY 1973. Bureau of Foods, Washington, D.C.
35. United States Department of Health, Education and Welfare. 1969. Report of the Secretary's Commission on Pesticides and their Relationship to Environmental Health. U.S. Govt. Print. Off., Washington, D.C.
36. Hindin, E., D.S. May, G.H. Dunstan. 1966. Distribution of insecticides sprayed by airplane on an irrigated corn plot. pp. 132-45 *in* Organic Pesticides in the Environment. Amer. Chem. Soc. Publ.
37. Ware, G.W., W.P. Cahill, P.D. Gerhardt, J.M. Witt. 1970. Pesticide drift. IV. On-target deposits from aerial application of insecticides. J. Econ. Entomol. 63:1982-83.
38. Buroyne, W.E. and N.B. Akesson. 1971. The aircraft as a tool in large-scale vector control programs. Agr. Aviat. 13:12-23.
39. Yates, W.E. and N.B. Akesson. 1973. Reducing pesticide chemical drift. chap. 7 *in* Pesticide Formulations. W. van Valkenburg, ed. Marcel Dekker, New York.
40. United States Department of Agriculture. 1975. Farmers' use of pesticides in 1971...expenditures. Econ. Res. Serv., Agr. Econ. Rep. No. 296.
41. Edwards, C.A. (ed.) 1973. Environmental Pollution by Pesticides. Plenum, London.
42. Heady, E.O., H.C. Madsen, K.J. Nicol, S.H. Hargrove. 1972. Future water and land use: effects of selected public agricultural and irrigation policies on water demand and land use. *in* Rep. Cen. Agric. Rural Dev., IA State Univ. Sci. Technol. NTIS, Springfield, Virginia.
43. Pimentel, D. and C. Shoemaker. 1974. An economic and land-use model for reducing insecticides on cotton and corn. Environ. Entomol. 3:10-20.
44. Newsom, L.D. 1975. Pest management: concept to practice. pp. 257-77 *in* Insects, Science, and Society. Academic Press, New York.
45. Bottrell, D.G. and P.L. Adkisson. 1977. Cotton insect pest management. Annu. Rev. Entomol. 22:451-81.
46. Carson, R. 1962. Silent Spring. Fawcett, Greenwich, Connecticut.
47. Shotwell, T.K. 1975. Patterns in federal regulation of pesticides. Pest Control 43(1):16-19, 22.
48. Food and Drug Administration. 1972. Revision of defect action levels for spinach. In-house memorandum, 14 December. U.S. Dept. HEW, Washington, D.C.
49. ————. 1974. Current levels for natural or unavoidable defects in food for human use that present no health hazard. Dept. HEW, PHS, Rockville, Maryland. (Fifth revision).

DISCUSSION:

R. WHETSTONE: You mentioned the effect of 2,4-D on corn. Do
the changes that occur in corn improve the food or feed quality of corn for man or animals?
D. PIMENTEL: All we checked was the N or protein content of
the treated corn. The protein N was significantly higher in
treated corn. Assuming all other nutrients in the corn remained the same, we might expect this treated corn with a
higher protein content to be better feed for cattle. This,
however, is just speculation. We don't know.
E. GLASS: When you state that as insecticide use has increased
over the years, losses have also increased, the inference is
that growers are actually using pesticide and still losing
more of the crops they are treating. Doesn't the loss
actually come on the crops where insecticides are *not* used?
In corn and apples, there's less damage now that insecticides
are used.
D. PIMENTEL: Corn losses have increased with the use of
insecticides according to USDA data. With apples, losses
have remained about the same. Concerning apples, there have
been quality changes that have influenced the loss estimates.
M. SAVOS: You are implying that we could do without pesticides
in agriculture; but how many more people would have to work
in agriculture if we did?
D. PIMENTEL: The number of workers would increase significantly, especially for weed control. Replacing herbicides for
weed control by mechanical means alone would at least double
or triple labor needs for weed control in row crops.
H. OLKOWSKI: Were your estimates of the increases in crop
losses due to not using pesticides based on using only current
techniques, or more sophisticated alternative controls?
D. PIMENTEL: Only readily available alternatives were used in
these estimates, such as mechanical control for herbicides—
but I want to emphasize that these are only estimates! In
general, we believe that they are overestimates. In apples,
for example, in our calculations the combined loss from
insects, diseases, and weeds were approximately 130% and
obviously this is an overestimate. We did not know how to
deal with this overlap problem.
E. DECK: I question your figures on the 65% aerial application,
and that 50 to 75% of the pesticides sprayed by air are lost
to the atmosphere. This is true only of the worst possible
dust formulations. With modern application techniques and
with liquid formulations, about 85-95% reaches the crop.
D. PIMENTEL: That 65% of all pesticides are applied by aircraft is based on USDA data. All the data that I have seen in
the literature indicate high loss of aircraft-applied

pesticides. In a recent study we conducted in Central America, 50-60% was lost under the ideal recommended application conditions. Other recent studies in California and in Arizona have substantiated that 50% to 75% of the pesticides applied by aircraft do not land in the target area. I don't know where your 15% figure came from. I would like to see your references.

T. CLARKE: Regarding the high cost of registering aphids, EPA retains the right to register or not any substance, with a mandate broad enough to include aphids.

E. JOHNSON: Legally we have that right, but we already have a draft document prepared that differentiates among the biologicals on an item by item basis.

R. TROETSCHLER: Do you see a reversal in the trend toward lowering tolerance levels for insects like aphids in food?

D. PIMENTEL: I would hope so, but I am dubious of the political feasibility of such an action by FDA.

T. GRUMBLY: You have to understand that we at FDA are under pressure from consumer groups to insist on 100% protection from materials that are commonly believed to be "undesirable" in food.

T. HULLAR: What would be some strategies that could take into account all these factors, and bring the $2:1 benefit/cost ratio of using pesticides back up, including external effects?

D. PIMENTEL: One would be to use pesticides only when needed. This can be done by monitoring pest and natural enemy populations. We also need more knowledge of economic aspects of pesticide use; probably 98% of present recommendations are not supported by such benefit/cost analyses. A great step forward was made in reducing external costs in the changeover from the persistent to the less persistent pesticides. These three steps can and will help to improve the benefit/cost ratio in pesticide use.

A LOOK AT U.S. AGRICULTURE IN 2000

Roger W. Strohbehn

Natural Resource Economics Division
Economic Research Service
United States Department of Agriculture
Washington, D.C.

U.S. agriculture, supported by the USDA/land-grant university research system, has provided the nation with abundant
food supplies at moderate prices. Simultaneously, it has contributed in a significant way to meeting world food needs and
maintaining a favorable balance of trade situation through a
high level of farm exports. About 30% of our crop production
is exported. And, on the average, U.S. consumers only use
17% of their disposable income for food.

Throughout the 1950s and 1960s, U.S. agriculture was
viewed as a cornucopia, and attention was focused on how to
hold down production to maintain a balance between supply and
demand. We became complacent about the ability of the agricultural sector to respond to long-run increases in the demand
for agricultural commodities. This complacency, however, was
shattered in the 1970s.

A series of independent changes occurred that raised new
concerns about the productive capacity of U.S. agriculture to
meet domestic and world food demand of the future. Environmental protection rules restrict farmer's decisions; weather
conditions have been less favorable; the development of new
agricultural technology appears to have slowed; energy prices
have skyrocketed, affecting both fertilizer and fuel costs;
import policies of several nations have been liberalized; and
third world nations are calling for a world food policy to
reduce starvation and malnutrition. The conventional wisdom
about unlimited productive capacity of agriculture has been
seriously challenged. The supply of agricultural commodities
may not be as easily expanded as in the past; and the demand
for commodities may grow at a faster rate than previously
anticipated.

73

At this Symposium you will be exploring alternative
strategies for pest control. The ultimate question is: "How
can U.S. agriculture manage its productive resources to maintain food supplies for domestic and world markets without
adversely affecting the environment?" Or stated a bit dif-
ferently, "Does agriculture have the capacity to meet growing
domestic and world food demands in the coming years without
environmental insults?"

To provide some insight into this question, I would like
to discuss with you some of the findings of a recent study
USDA completed for the Water Resources Council. The primary
purpose of this study was to estimate agricultural water needs
for the year 2000, relate them to available supplies, and
determine potential future water shortages.

A comprehensive study was made to project the behavior
of U.S. agriculture in 2000 and provide a basis for estimating
water needs and evaluating impacts of water shortages. This
study provided a wealth of information about the capacity of
American agriculture. Several different scenarios of the
situation in 2000 were considered, including an environmental
protection scenario. Even though alternative pest management
strategies were not included in the scenarios, our findings
do provide background for considering the economic and social
implications of pest management strategies.

The field of futures analysis is filled with land mines.
The future is seen only dimly. One only has to go back in
time and try to visualize our world of 1977 from the perspective of 1952 conditions to appreciate the difficulty of looking ahead to the year 2000. Who in 1952 foresaw the environmental movement except perhaps Rachel Carson and Barry
Commoner? Who in 1952 would have predicted that corn yields
would increase from 42 bushels per acre to 97 bushels per acre
by 1972? Who in 1952 would have predicted substantial grain
trade with Russia and China by 1973? As a statistician friend
once said, "things unknown are highly uncertain."

In spite of the uncertainty of future events, we made a
series of assumptions to provide a basis for estimating how
agriculture might look in 2000. Please note that we were not
forecasting or predicting what will happen in 2000, but rather
simply indicating what *could* happen if the assumptions we made
are a reasonable description of future conditions.

ANALYTIC BACKGROUND

To provide some background and understanding of how we
arrived at our estimates for 2000, I will describe how the
projections were made.

We made projections for an "historic trend benchmark" in
the year 2000 that assumed a continuation of growth of the
economy, population, per capita consumption of food, exports,
crop yields and conversion of agricultural land to urban uses.
We estimated output, land use, land development, earnings,
employment, and soil loss associated with the production of
20 crops and 7 livestock categories that represent the bulk
of agricultural production of the United States. In addition
to the benchmark projection, we made a series of alternative
future projections to describe the consequences of following
different public policies relating to exports, resource development, and environmental protection. In this paper, I will
focus on the benchmark projection and the high-level export
alternative.

We used a national linear programming model to project
how farmers in various regions of the United States would use
their land, labor, and capital resources to meet domestic and
export demand in the future. The nation was divided into 105
producing areas in which crop and livestock activities were
defined in terms of the production costs and yields relevant
to each of the producing areas. Pesticide applications were
reflected in the crop budgets; however, they were not included
in a manner that allowed alternative applications levels to be
evaluated.

In addition to the producing areas, we designated 28 market regions in the United States. We then estimated domestic
and export demand for food and fiber in each region based on
its projected population and the cost of shipping commodities
to ports of export.

The producing areas and the market regions were linked by
a transportation network that reflects the cost of shipping
commodities from producing areas to consumption and export
centers. Thus, the linear programming model was designed to
simulate the manner in which farmers in different regions
compete with each other to provide the level of output demanded for domestic consumption and exports.

Linear programming models formulated for this study operate on economic efficiency criteria that seek to achieve a
specified output at least cost. By only using economic criteria to identify where production can be obtained at the
lowest cost, an implicit assumption is made that farmers will
quickly adjust to the most efficient locations of production
and only use the quantity of land and other inputs to achieve
the market level output. Farmers, of course, do not make
immediate adjustments due to immobility of land and farm
operator labor and a variety of other noneconomic reasons. In
recognition of the slow adjustment process, we placed constraints on the model to retain a proportion of historic

production in each producing area. Each area had to use at
least 40% of the 1969 acreage devoted to crops grown in the
respective producing area by 2000. In addition, each area
could not expand its production by more than 600%.

In a nutshell, the approach used to make the projections
of the agricultural situation in 2000 was to bring together
systematically information about the resource availability
and production potential of all regions of the United States
to meet anticipated commodity demand levels in an objective
manner.

AGRICULTURE IN 2000

What might be the features and characteristics of agriculture in 2000? Specific assumptions selected to depict a
"benchmark-trend" future in 2000 were as follows—
1. population growth of 0.84% per year resulting in a
   U.S. population of 264 million;
2. moderate growth in crop yields—1.3% annual growth
   compared with the 2.8% annual growth from 1950 to
   1975;
3. cropland base of 425 million acres plus potential
   development of 23 million acres by drainage and
   irrigation;
4. moderate growth in export demand as viewed by international trade experts in 1974 (USDA has revised its
   long-range export estimates upward since this study
   was completed);
5. soil conservation practices applied by farmers to
   limit soil loss to 2 "T" to achieve environmental
   quality goals[1].

Land Use

By the year 2000, 372 million acres of land is projected
to be used for crop production. This is about 50 million
acres more cropland than was reported in our 1971-73 base
period, but only about 20 million acres more than were used
in 1976. To a large extent, the increase in domestic consumption and exports will be counterbalanced by the projected
increase in agricultural productivity. Substantial cropland
developments are projected to occur by 2000. Clearing and/or

---

[1]*"T" value reflects the maximum annual soil loss that can
occur on a given acre of land and still maintain its productivity.*

draining of wet soils in the humid eastern states would add
10.5 million acres. Irrigation development would occur on
6.7 million acres while nearly an equal number of irrigated
acres would revert to dry land farming.

The expansion of crop production is not expected to occur
uniformly across the nation. Regions that could be major
gainers in expanding their acreage devoted to crop production
(i.e., increase by 25% or more) include South Atlantic Gulf,
Great Lakes, Ohio, Tennessee, Arkansas-White-Red, Texas Gulf,
and the Upper Colorado. In contrast, regions that could
remain constant or experience a reduction in cropland include
the Lower Mississippi, Rio Grande, Columbia North Pacific,
and California. All other regions are likely to experience
moderate expansion in cropland acres by 2000 (see Table 1).

Large acreages of currently arable but uncropped land
will still exist in future years. According to our projections, U.S. agriculture would still have a reserve capacity
of 53.7 million acres of unused cropland that could be readily
put into crop production. To a large extent these acres
represent less productive, marginal lands. Nevertheless, they
are part of the acreage that farmers consider cropland and
deem suitable for crop production under appropriate land
management practices.

Water Use

Water availability will continue to be a major factor
affecting agricultural production in several western regions
(fig. 1). At the present time, there are about 37 million
acres of harvested irrigated cropland in the western regions,
on which 103 million acre-feet of water are applied. By
2000, depletions in groundwater supplies, primarily in western
Texas, will offset irrigation development elsewhere and result
in a net increase of only 800,000 acres. The Upper and Lower
Colorado regions plus the Great Basin and Rio Grande regions
will utilize nearly all of the available agricultural water
supplies in their respective regions. In other regions water
supplies appear adequate to meet the nation's projected level
of food and fiber demand in 2000.

In addition to the benchmark-trend scenario, we also
developed a high-export scenario.

High-Export Scenario

The high-export scenario was based on a continuation of
the commodity export expansion experienced in the 1970s.

Table 1. Regional Cropland Use for 1971-73 and Alternative Futures in 2000

Water Regions	1971-73		2000 Benchmark		2000 High-Export	
	Cropland used	Reserve capacity	Cropland used	Reserve capacity	Cropland used	Reserve capacity
	million acres					
New England	1.7	1.1	1.9	0.7	2.3	0.3
Mid Atlantic	9.2	2.5	10.5	0.6	11.2	0.4
S. Atlantic Gulf	17.5	7.8	23.7	3.9	29.8	1.1
Great Lakes	18.7	6.2	24.0	1.3	24.7	0.7
Ohio	24.8	8.1	31.0	2.0	31.7	1.5
Tennessee	2.5	1.9	3.2	1.3	3.4	1.0
Upper Mississippi	54.4	9.7	61.8	2.9	62.4	2.3
Lower Mississippi	18.2	2.2	18.1	4.4	22.7	0.4
Souris-Red-Rainy	16.0	4.8	19.5	1.4	20.5	0.4
Missouri	83.5	22.2	89.6	15.5	96.2	9.2
Arkansas-White-Red	29.7	14.8	38.1	7.0	41.5	3.9
Texas Gulf	13.9	11.1	17.9	6.6	21.7	3.2
Rio Grande	2.4	0.5	2.1	0.7	2.2	0.6
Upper Colorado	1.4	0.6	2.0	0.1	2.0	0.1
Lower Colorado	1.2	0.3	1.3	0.1	1.3	0.1
Great Basin	2.0	0.6	2.3	0.5	2.4	0.3
Columbia N. Pacific	16.1	3.1	15.5	3.5	17.6	1.3
California S. Pacific	9.1	1.8	9.1	1.2	9.5	0.8
United States	322.3	99.3	371.6	53.7	403.1	27.6

78

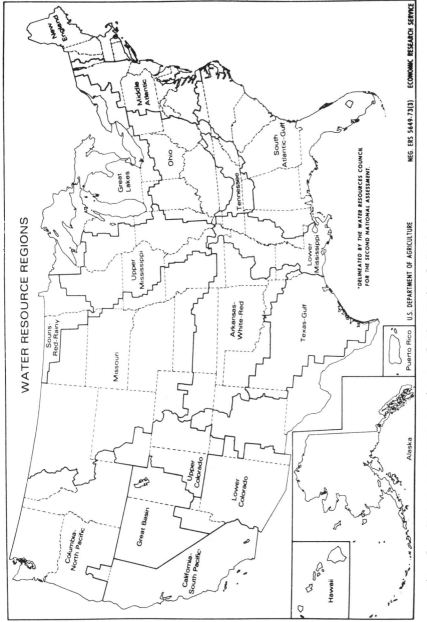

WATER RESOURCE REGIONS

*DELINEATED BY THE WATER RESOURCES COUNCIL FOR THE SECOND NATIONAL ASSESSMENT.

U.S. DEPARTMENT OF AGRICULTURE   NEG. ERS 5649-73(3)   ECONOMIC RESEARCH SERVICE

Fig. 1. Water resource regions of the United States.

Under this scenario corn exports would be 160% above the
1973-75 export levels. Soybean exports would be 250% greater
than in 1973-75 and wheat exports would be up by 30%.
If this level of world demand were realized in 2000, 403
million acres would have to be used for crops and unused cropland would fall to 28 million acres. Two million more acres
would need to be irrigated and 16 million acres of uncropped
wet soils would need to be drained. Thus, agriculture could
meet this call to production and still have a small margin of
reserve capacity.

IMPLICATION FOR INTEGRATED PEST MANAGEMENT

This preview of U.S. agriculture in the year 2000 reveals
a fairly bright future. Resources appear adequate to meet
part of the rising world food needs even if demand continues
to grow at the rapid rate experienced in the last five years.
This assessment, however, should be tempered to reflect uncertainty about long-range weather cycles, rising energy
costs, and the slow response of nonfarm landowners to shift
land into crop production. Each of these factors could have
a dampening effect on agricultural productive capacity in
2000. The agricultural cornucopia we have known in the past
may not be quite as limitless as we believed. Through careful
management, U.S. farmers should be able to provide an abundant
supply of food and fiber for domestic and foreign markets.
Improvements in water quality of the nation's streams and
lakes would also occur through greater adoption of "best
management" soil conservation practices.
What implications can be drawn from this review for integrated pest management strategies? The production and cost
coefficients used in the study assumed a continuation of the
current package of pest control chemicals. Due to environmental and health consequences, many of these chemicals may
not be available in 2000. If biological and cultural controls
are less effective than chemicals in controlling the target
pest, the projected yield increase could be lower than anticipated in this analysis. This, of course, would imply the
need for more acreage to meet the future commodity demand
levels, and the reserve capacity of U.S. agriculture would be
reduced. Consequently, highly effective biological controls
will be needed to maintain a reserve capacity that enables
agriculture to respond to rising world markets.
If integrated pest management systems raise production
costs, this could translate into higher commodity prices,
with a consequent reduction in market demand. Less land,
water, and other resources would be required to meet the lower

demand level. Given the competitive nature of farmers, however, they might absorb the higher production costs, resulting
in a reduction in net farm income and lower land values.

In all cases, the challenge facing scientists and educa-
tors is to develop biological-cultural pest management systems
that will effectively and efficiently replace the chemicals
that have been discovered to have undesirable environmental
side effects.

DISCUSSION

UNIDENTIFIED: In Kansas we may run out of the energy needed
to pump water before we run out of water.

R. STROHBEHN: We did not take into account variations in
energy cost, which would affect the analysis.

W. LOCKERETZ: Have you considered how accurate previous USDA
projections have proved to be?

R. STROHBEHN: We are not trying to predict the future. Rather
we are trying to see what the consequences would be of various
policies and programs.

UNIDENTIFIED: Could possible effects of home gardens for urban
food production be included in your model, to see their
effects on exports, for example?

R. STROHBEHN: This could be built into our models, but probably would not have a significant effect on exports. The
acreages of food grown for home consumption are small compared
with, say, those for corn, soybeans, wheat, and other
commodities that are exported.

A. ASPELIN: How did the environmental scenario contrast with
the one you have presented?

R. STROHBEHN: This model limited the amount of water withdrawn
from each river, allowed no reclamation of wetlands, and
required the recycling of animal waste to the land. These
restrictions severely taxed the capacity of agriculture to
meet our commodity demands. We could meet those demands, but
only by using practically every available acre of land.

Part III
Case Studies of Pest Management

ALFALFA WEEVIL PEST MANAGEMENT SYSTEM FOR ALFALFA

Edward J. Armbrust

Illinois Natural History Survey and
Illinois Agricultural Experiment Station
Urbana, Illinois

INTRODUCTION

Alfalfa is the world's most valuable cultivated forage crop, and it is recognized as providing the best food value for all classes of livestock. In the United States there are nearly 30 million acres of alfalfa, which provide more than 50% of the total hay crop. Alfalfa is exceeded in total acreage by corn, wheat, and soybeans, but its protein production potential per acre far exceeds that of each of these three crops. Because alfalfa is a nationally grown crop that is adapted to a wide variety of geographic and climatic conditions, a research project[1] was begun in seven areas representing different agricultural and biotic regions. Each of these regions has its own unique agroecosystem and problems, but each contains certain alfalfa insect pests common to the other. The entire scope of problems is represented by those in California, Utah, Nebraska, Illinois, Kentucky, Virginia, and New York.

[1] *Integrated Pest Management Project (IPM), the Principles, Strategies, and Tactics of Pest Population Regulation and Control in Major Crop Ecosystems. This research was supported in part by the Illinois Natural History Survey, the Illinois Agricultural Experiment Station, National Science Foundation, and the U.S. Environmental Protection Agency, through a grant (NSF GB 34718) to the University of California. The findings, opinions and recommendations expressed herein are those of the author and not necessarily those of the University of California, the National Science Foundation, or the Environmental Protection Agency.*

A principal investigator was designated for each region to coordinate the team research effort at that respective institution. At the beginning of this project, avenues of communication and representation were established with various other alfalfa research groups on both a national and regional scale. In several instances, research was coordinated with outside projects. Several states that were not funded by this project supplied field data for some of the studies reported herein. This coordination and exchange of data assisted in making this a successful and productive project.

The alfalfa ecosystem is unique among field crops in that it is a relatively long-lasting, well-established perennial system. This makes it of prime importance ecologically. It is further unique in combining long- and short-term cycles of drastic environmental change with a mutualistic soil-plant relationship, and levels of resistance to certain pests, which permit great management flexibility.

Since the early 1940s, the major method of insect pest control in crop ecosystems has been the use of chemical insecticides against the damaging stages of a pest at the time of population outbreaks. This unilateral approach has often produced disastrous side effects on nontarget organisms and the environment. Because of the undesirable side effects of this method of insect control, many workers have strongly advocated a more fundamental approach—one that considers the basic relationships among host plant, pest populations, related abiotic and biotic factors, and the combined management practices used to produce the crop.

The main objective of the Integrated Pest Management Project for alfalfa has been to gain a better understanding of the behavior of the alfalfa ecosystem and develop multidisciplinary insect pest management programs for alfalfa that will improve production efficiency and environmental quality. The following subobjectives have assisted in fulfilling this goal: (1) to ascertain the ecological interactions of the components of the alfalfa ecosystem, (2) to develop simulation models of the alfalfa ecosystem, (3) to ascertain the basic biological parameter values necessary to quantify the above models, and (4) to verify and field test the models.

The alfalfa weevil, *Hypera postica* (Gyllenhal), and the closely related species, the Egyptian alfalfa weevil, *Hypera brunneipennis* (Boheman), have been the most important pests of alfalfa in the United States. The alfalfa weevil was first discovered near Salt Lake City, Utah in 1904, and for nearly 50 years it remained confined to 12 western states. However, in 1952 it was discovered in Maryland and from there it spread rapidly throughout the remainder of the United States.

The Egyptian alfalfa weevil was first discovered in the United States in 1939 and it has remained confined mainly to California and Arizona.

Even though one might consider the alfalfa weevil complex as a single national alfalfa pest problem, various geographic regions have their own unique problems because the weevils' biologies and control requirements are extremely dependent on climatic factors. Because the weevil complex, and more specifically *H. postica* and its widely distributed parasite, *Bathyplectes curculionis* (Thomson), is of concern to growers on a national scale, it has been the key insect pest under investigation in this project. Other insects, both pest and natural enemy species, were also considered in the total arthropod complex on alfalfa, and whatever knowledge could be gained from specific studies on them has been incorporated into the program. Because of the diversity of alfalfa production and management and the biological differences in the weevil complex in relation to crop development in the specific region, many aspects of the research were conducted in each of several regions.

Because there is a vast amount of available information on the biology, ecology, and control of alfalfa insect pests, an immediate effort was made to search, categorize, analyze, and utilize the existing literature and unpublished data. By the end of the present project there will be published, indexed bibliographies on the spotted alfalfa aphid (1), pea aphid (2), *Sitona* species (3), potato leafhopper, weevil complex, and weevil parasites. This represents over 9,000 literature references of which two-thirds have been indexed and entered into a computerized literature retrieval system[2]. This accomplishment has been of valuable assistance to the biologists, as well as those involved in modeling efforts. In many instances it was found that parasites had been established, cultural practices and timing of insecticide applications had been investigated, and considerable effort had been put forth in developing resistant varieties. These data have been integrated into a system that uses each factor in a complementary fashion.

There are over 2,000 literature citations dealing with the weevil complex and its parasites, but it was apparent that certain aspects of alfalfa weevil and parasite biology needed special research attention. The modeling effort helped to distill out these unknowns and point to specific research needs. Especially needed was a better understanding

[2]*Illinois Natural History Survey and INTSOY, 163 Natural Resources Building, Urbana, Illinois 61801*

of the mortality factors influencing fluctuations in the
insect pest and natural enemy populations in order to
develop a better insect management system for alfalfa, a
system that depends increasingly on improved insights gained
from the systems approach.

ALFALFA WEEVIL BIOLOGY

Alfalfa weevil larvae have green bodies and black heads
and are 3/8 in. long when full grown. They feed on alfalfa
plants for 3 or 4 weeks in the spring. During this time they
shed their skins 3 times. When full grown, they spin silken
cocoons on the plants, within the curl of fallen dead leaves,
or in litter on the ground. They change into adults in 1 or
2 weeks. After feeding for a short time in the spring, these
adults fly to protected areas and enter a resting period. In
many areas, most of the adults will return to the alfalfa
fields in late summer and early fall. In some northern areas
the adults return to the alfalfa the following spring.

In southern areas, if temperatures permit, the weevils
will lay eggs throughout the fall and winter as well as into
the spring. Some of the eggs will begin to hatch about the
time alfalfa is beginning its spring growth. In the more
northerly regions, the larger number of eggs are laid in
the spring. By the time the larvae emerge, the alfalfa is
6 to 10 in. tall and can tolerate more weevils than the
southern crop.

Because population peaks vary from year to year, it is
difficult to predict when spraying will be necessary.
Existing methods of determining when to spray (such as percent
tip feeding) are often confusing and do not consider crop
height or weevil numbers.

CONTROL METHODS AVAILABLE

Although insecticidal control has been the most widely
used method, two other methods are common. One is to
manipulate the timing of the first harvest in the spring.
After considering many factors such as numbers of weevil
larvae, plant growth, and prevailing weather conditions,
proper timing of the cutting date can achieve the same effect
as the application of an insecticide. The other method
involves biological control agents, such as parasites and
predators. One of the most successful biocontrol agents is
a small parasitic wasp, *Bathyplectes curculionis*. This wasp
lays its eggs inside young weevil larvae. The wasp larvae

develop inside the weevil larvae, and when they have satisfied their needs, they kill their hosts. Other parasites of the larval stage include: *Bathyplectes anurus, Bathyplectes stenostigma, Tetrastichus insertus;* of the egg stage, *Patasson luna;* and of the adult stage, *Microctonus colesi, Microctonus aethiopoides.*

A high level of resistance in alfalfa to *Hypera* spp. has not been found, but varieties with low levels of resistance have been developed (4). These previously developed varieties are of the dormant type and were tested exclusively against *H. postica.* Summers and Lehman (5) demonstrated that resistance can be found in nondormant type alfalfas and that varieties developed for resistance to *H. postica* also resist *H. brunneipennis* (Table 1).

Table 1. Larval Populations, Damage Ratings, Tolerance Ratings, and Leaf Weight in Seven Alfalfa Cultivars, Fresno County, California (5).

Cultivar	Mean no. larvae sweep[a]	Damage rating[a,b] whole plot	single stem	Tolerance rating[a,c]	Dry wt. (mg) of leaves/ cm stem[a]
Team	40.0ab	3.6a	3.5a	4.84a	13.81a
Weevilcheck	38.9ab	3.5a	3.6a	4.26b	14.28a
UC 73	35.9a	3.3a	3.3a	4.27b	14.07a
UC Cargo	59.7bc	5.3b	5.7a	2.84d	10.58b
UC Salton	60.3bc	6.0b	5.2b	3.45c	11.68b
UC 67	69.8c	5.5b	4.7b	3.48c	11.71b
Moapa	65.4c	5.0b	5.4b	3.60c	11.92b

[a]Means followed by the same letter(s) are not significantly different at the 5% level of probability. Duncan's multiple range test.
[b]1 = no defoliation, 10 = complete defoliation.
[c]Higher values indicate greater tolerance to feeding damage.

These data verify the importance of the tolerance component in resistance to *Hypera* spp. but also indicate the possible presence of other components of resistance.

PURPOSE OF ALFALFA WEEVIL PEST MANAGEMENT

The principal reason for a pest management program is that the three control methods are interrelated. For example, insecticides kill parasites and predators as well as alfalfa

weevils. Harvesting alfalfa when many of the weevil larvae
are parasitized will also reduce the parasite population. To
help understand how the three methods of control work together, a computer-based mathematical model was developed. This
model simulates field conditions. The pest management program resulted from laboratory analysis of the model, followed
by extensive field trials. The values in published charts
(6) were taken directly from these trials. The charts are
designed to explain the model without requiring the use of
a computer.

This program assumes proper soil pH, adequate fertility,
and that the alfalfa is not under stress from drouth, disease,
or other factors. If these conditions are not met, it is
possible that excessive feeding injury will occur.

STANDARDIZATION AND IMPROVEMENT OF RESEARCH TECHNIQUES

In order for researchers to be able to compare results
from alfalfa weevil field studies, for each life stage,
standardized research and sampling techniques that were based
on a unit area basis were agreed upon. It was further
agreed that the sweep net should not be used for population
studies until sweep net catch could be related to absolute
densities.

Where necessary, research results were obtained to
refine current methods of sampling for each life stage,
and it was soon discovered that sampling intervals would best
be based on physiological time as measured by degree days.

ADULT ALFALFA WEEVIL SAMPLING

Estimates of absolute density (weevils per square foot)
for adult alfalfa weevils have been notoriously difficult to
obtain. Even under the best conditions about one man-hour is
needed in the field plus 3 days of extraction time and use of
20 Berlese funnels in the laboratory to estimate adult
density for one field. A more typical case would ·require
about 3 to 4 times this effort.

A study was undertaken in Illinois to obtain information
about the relationship between sweep net catch and absolute
density. It was hypothesized that the ratio (M) of absolute
density to catch per sweep would depend largely on alfalfa
height, temperature, wind speed, relative humidity, and solar
radiation intensity. Data were gathered between April 1973
and November 1975 from several fields in Illinois and Indiana,
resulting in 29 data sets.

Multiple regression analysis was used to obtain an

equation for predicting the ratio M from environmental conditions. The best equation, based on analysis to date is

$$M = 0.2 + 10^{7.3595 - .001343(T + 3L)^2 - .3744H + .0058TH}$$

where T is temperature in °F, H is crop height in centimeters, and L is solar radiation in Langley units (cal/cm^2/min). Use of this conversion equation is facilitated by using tables, with which one can estimate M in about 15 seconds.

This equation has not been field tested using independent data, and until that time it is premature to place much confidence in its predictions. It is expected that absolute density can be estimated within 20% of the true mean 90% of the time, based on 200 to 500 sweeps (depending on catch per sweep), and that about 15 to 30 minutes will be required in each field.

COLLECTION OF FIELD SAMPLES FOR LIFE TABLE CONSTRUCTION

Using standard sampling techniques for each life stage of the weevil, population data for the weevil complex and for *B. curculionis* have been collected over a period of 3-4 years from one or more locations in Kentucky, California, Illinois, Indiana, Virginia, New York, Ohio, Utah, Michigan, and Nebraska. These data have been extremely valuable for determining age specific survival rates and key mortality factors, quantifying the effects of population dynamics and impact of parasites on population regulation, and field validation of models.

The life tables were based on 1 ft^2 samples of alfalfa collected at random within each field. Eight age intervals were used to trace the course of each generation. Graphic key factor analysis was used to study the factors responsible for changes in population density. By this method, the killing power, or k-value, of each mortality factor was estimated by taking the difference between the logarithm of population density before and after its action. Using these values, the data from Kentucky were plotted for 11 life tables. The results indicated that k_g, mortality of summer adults (to emigration), followed the same fluctuating course as did changes in generation mortality (K). Thus total generation mortality was dominated by the magnitude of summer adult mortality (cause not specific) and this represented the key factor of importance in explaining generation survival of *H. postica* on alfalfa. At the same time, mortality of larvae due specifically to parasitism tended to compensate for the changes in adult mortality and

reduce generation-to-generation variation in K, and thus
contributed a density dependent regulatory component[3].

QUANTITATIVE BIOLOGY STUDIES

   In the construction of insect models for the alfalfa
weevil and *B. curculionis,* it soon became apparent that there
were certain research areas needing special attention for both
alfalfa weevil and *B. curculionis*. These were: diapause
rate, diapause duration and termination, adult survival,
oviposition rate, and dispersal between habitats. A detailed
quantitative understanding of these features as affected by
different, or changed environmental conditions was needed.
   A literature search at the beginning of this project
revealed few detailed studies concerning the biology of
the parasites and the impact of insecticides on them. Even
more limited were studies dealing with predators in alfalfa
and their impact on the alfalfa weevil or its parasites.
Hundreds of species of organisms are commonly found in
alfalfa, and the role of many is as yet poorly understood.
In these studies, to avoid disrupting whatever natural
stability may exist in this ecosystem, the system was dis-
turbed as little as possible. Thus, investigations of the
natural control complex in relatively undisturbed situations
became an essential part of this project. There has been
wide variation, however, in the emphasis of effort on
specific members of the natural enemy complex.

MORTALITY FACTORS AFFECTING *B. CURCULIONIS*

   Applications of pesticides in the alfalfa ecosystem
interrupt the normal relationship between alfalfa and its
associated arthropod fauna. Cocoons of *B. curculionis* can
be found in the litter for a period of 1-2 weeks for nondiapausing individuals and 10-11 months for diapausing ones.
Bartell et al. (7) demonstrated that the construction of the
cocoon, and the metabolic state of the individual within it,
influences the penetration of insecticides.
   Because the cocoons of diapausing *B. curculionis* remain
in the litter from one spring to the next they are vulnerable
to insecticides over a much longer period of time, and to a
variety of other factors such as weather, predators,

[3]*M.A. Latheef and B.C. Pass, University of Kentucky,
Lexington, unpublished data*

parasites, pathogens, cultural practices, etc. Cherry and
Armbrust (8-10) found that invertebrate predation caused the
heaviest mortality in these diapausing larvae. Specific
predators were identified by exposing *B. curculionis* larvae
to various surface-dwelling invertebrates found in alfalfa
fields. In addition, field plantings of parasite larvae in
modified screen mesh cages were used to determine the size of
the predators involved and also to determine if litter
density affected rates of predation or the species involved.

Results from feeding studies showed that spiders,
Cicindelidae, Formicidae, and small species of Staphylinidae
did not prey on *B. curculionis* larvae in cocoons. The two
groups of predators that did consume them were field crickets,
Gryllus pennsylvanicus Burm., and various species of Cara-
bidae. The predators of the parasite larvae planted in the
field were mainly insects of moderate size, and their use of
B. curculionis was not significantly affected by litter
density. The greatest number of total predators (*G. pennsylvanicus* plus carabids) caught/day/pitfall trap, and the
greatest predation on field-planted *B. curculionis* larvae,
occurred concurrently during September and October. These
data suggest that predation during fall (September and
October) may be significant in reducing field populations of
diapausing parasite larvae. Based on feeding studies and
pitfall trap catches, *G. pennsylvanicus* and the carabids,
Abacidus permundus (Say), *Evarthrus sodalis* LeConte, *Harpalus
pennsylvanicus* DeGeer, and *Scarites subterraneus* Fab., were
the most significant specific predators on *B. curculionis*
larvae. These and other invertebrate predators cause a far
greater mortality to diapausing *B. curculionis* than the
combined effects of weather, hyperparasites, diseases, and
the insecticide usage.

ECONOMIC THRESHOLDS AND COMPENSATION OF ALFALFA TO INSECT
DAMAGE

Early in this project some states, especially California
(11), began economic injury studies for the weevil complex
and certain other participating institutions cooperated with
researchers at Purdue University to determine these levels
under various biological and geographical conditions (12).
Also, in order to evaluate management practices and control
measures, quantitative knowledge was required of the effects
of insect defoliation on yield and quality. In New York,
Liu and Fick (13) compared yield and quality in systems cut
twice and three times during the season, where insect pests
were controlled and not controlled. The greatest effect for

a single harvest was in the 2nd cutting of the 3-cut system
where feeding occurred in the early stages of regrowth.
Indirect plant responses to insect feeding may partially
compensate for direct losses or, on the other hand, cause
further reductions in yield and feed quality. Fick and Liu
(14) found many indirect effects of insect defoliation that
have to be considered in evaluating insect damage to alfalfa
and included in plant and insect models.

PRACTICAL ACHIEVEMENTS TOWARD IMPROVED WEEVIL CONTROL

Detailed descriptions of the population dynamics of the
alfalfa weevil and its parasite have made it possible to
develop management models. For example, defining the
phenology of the weevil and its parasite has led to adjustments in management practices to maximize parasite survival
and minimize weevil survival. Furthermore, the development
of a simulation model for alfalfa growth has added another
dimension to understanding the weevil/alfalfa interaction.
Predictions of plant growth and phenology should allow for
improved scheduling of management practices such as harvest.
In addition to the basic biology of the alfalfa plant,
weevil, and parasite, an understanding of such phenomena as
plant compensation to insect damage, disruption of ecological
balance in the alfalfa crop system, and the economics of the
weevil/alfalfa interaction will lead to a more realistic
structuring of the alfalfa ecosystem. The results suggest
new approaches to management of the alfalfa weevil. For
example, this past season, alfalfa researchers in cooperation
with Extension staff and alfalfa growers, continued an alfalfa
weevil management program in alfalfa fields in Illinois,
Kentucky, New York, Iowa, Utah, Virginia, and other
cooperating states.

HOW TO USE THE PROGRAM

To use this program the following things have to be done:
1. Calculate degree-day accumulation by recording daily
   high and low temperatures from January 1 until the end
   of the alfalfa weevil season in late spring.
2. Count the number of larvae on a 30-stem sample.
3. Measure the height of 10 stems from the original 30.
4. Refer to Recommendation Charts. They provide directions for the entire weevil season. They tell you
   either to resample, harvest early, or spray. They

also tell you when the weevil season is over and you
no longer need to sample.

Measuring and Recording Temperature

A record of daily high and low temperatures should be
kept from January 1 until the end of the alfalfa weevil
season. This information can be obtained from the daily news-
paper, local weather stations, radio or television, specialized county Extension systems, etc. Once the daily high and
low have been obtained, the next step is to convert this
information into degree-days from published tables.

The degree-days used in this program are based on a
developmental threshold of 48°F., since at temperatures lower
than this little or no weevil development takes place. (Note
that the degree-days used to calculate alfalfa weevil development are not the same as the degree-days quoted in weather
reports.) In areas where alfalfa weevils lay eggs in the fall
and winter, field sampling must begin when 200 degree-days
have accumulated since January 1. In areas with no fall or
winter egg laying, sampling need not begin until 400 degree-
days have accumulated.

Counting Larvae and Measuring Plant Height

Each 30-stem sample from a field should be taken in a
pattern that covers as much of the field as possible. This
is important because the level of infestation varies across a
field. For example, the problems are often worse on southern
slopes because these areas tend to be more protected during
the winter, and warm up sooner in the spring. Field edges
should be avoided because they give inaccurate samples. If
possible, stay at least 50 feet from the edges.

At 30 evenly spaced intervals, carefully pick an entire
stem (without dislodging any larvae) and place it in a 2- to
3-gallon container. Stems at each location must be selected
at random, and this can be done by picking the first stem the
hand touches. Next, beat the 30 stems vigorously against the
inside of the container for a few seconds. Transfer the
larvae to a shallow pan for counting and record the number
found. Randomly select 10 stems from the original 30 and
record their average length to the nearest inch.

This process requires 20 to 25 minutes for a 15- to 20-
acre field. In very large fields (40 acres or more), it may
be better to take two or more 30-stem samples and average the
results.

If the alfalfa was windrowed during harvest, the 30-stem samples should be picked from the windrow area whenever possible (after removal of the hay). If there are enough larvae on these stems to recommend spraying, it would be well to pick another 30 stems, avoiding windrow areas. If there are so few larvae on these stems that spraying is not recommended, spraying only the windrow areas will save on the cost of insecticide.

Sampling Frequency

Samples should be taken more frequently early in the season than toward the end of the season. Expect to visit a field an average of every 7 days during the weevil season. With an extremely early spring, a field could be sampled as many as 11 times or more.

A sample that is preceded by frost or beating rains can result in underestimation of population density. Numerous larvae may be found on the ground following these weather conditions. Although some larvae will probably fail to crawl up the plant, it is suggested that these kinds of fields be resampled the following day.

Each time an alfalfa field is sampled, Recommendation Charts must be consulted to determine if spraying is needed. A field must be resampled 100 degree-days after spraying to make sure the spray was effective.

Unusual Situations

Although the Recommendation Charts are designed to allow decision making in a routine fashion, unusual weather conditions may require a few modifications. With several alfalfa fields, unseasonable warm weather early in the season (such as during February) could make it difficult to finish sampling all fields within the prescribed degree-day ranges. In this case, starting or finishing 5 to 10 degree-days on either side of the range will not be detrimental. However, subsequent sampling should be adjusted to coincide with the next range on the chart. This situation will be rare. Usually the rate of weevil development and plant growth varies enough from field to field that, after the first sample of the season, all fields will rarely be sampled on the same day.

SUMMARY

   This program is offered as an alternative for deciding when to spray alfalfa weevil. Older methods are really no better than rules of thumb and are often confusing. As a result, treatment thresholds will vary greatly with the observer.

   Although the Alfalfa Weevil Pest Management Program has received extensive testing, it is being continually improved, and will receive further revision and refinement. The program offers the farmer, pest management consultant, dealer, and others methods for determining the timing of insecticide applications for control of the alfalfa weevil.

REFERENCES

1. Davis, D.W., M.P. Nichols, E.J. Armbrust. 1974. The literature of arthropods associated with alfalfa. 1. A bibliography of the spotted alfalfa aphid, *Therioaphis maculata* (Buckton) (Homoptera:Aphidae). Ill. Nat. Hist. Surv. Biol. Notes 87.
2. Harper, A.M., J.P. Miska, G.R. Manglitz, E.J. Armbrust, B.J. Irwin. 1977. The literature of arthropods associated with alfalfa. 3. A bibliography of the pea aphid, *Acrythosiphon pisum* (Harris) (Homoptera:Aphidae). Univ. Ill. Misc. Publ. (in press).
3. Morrison, W.P., B.C. Pass, M.P. Nichols, E.J. Armbrust. 1974. The literature of arthropods associated with alfalfa. II. A bibliography of the *Sitona* species (Coleoptera: Curculionidae). Ill. Nat. Hist. Surv. Biol. Notes 88.
4. Sorensen, E.L., M.C. Wilson, G.R. Manglitz. 1972. Breeding for insect resistance. pp. 371-90 *in* Alfalfa Science and Technology. C.H. Hanson, ed. Am. Soc. Agron., Madison, Wisconsin.
5. Summers, C.G. and W.F. Lehman. 1976. Evaluation of nondormant alfalfa cultivars for resistance to the Egyptian alfalfa weevil. J. Econ. Entomol. 69:29-34.
6. Wedberg, J.L., W.G. Ruesink, E.J. Armbrust, D.P. Bartell. 1977. Alfalfa weevil pest management program. Univ. Ill., Coll. Agr. Coop. Ext. Serv. Circ. No. 1136.
7. Bartell, D.P., J.R. Sanborn, K.A. Wood. 1976. Insecticide penetration of cocoons containing diapausing and nondiapausing *Bathyplectes curculionis*, an endoparasite of the alfalfa weevil. Environ. Entomol. 5:659-61.
8. Cherry, R.H. and E.J. Armbrust. 1975. Field survival of diapausing *Bathyplectes curculionis*, a parasite of the alfalfa weevils. Environ. Entomol. 4:931-34.

9. ———. 1977. Predators of *Bathyplectes curculionis*, a parasite of the alfalfa weevil. Entomophaga (in press).
10. Cherry, R.H., E.J. Armbrust, W.G. Ruesink. 1976. Lethal temperatures of diapausing *Bathyplectes curculionis* (Hymenoptera:Ichneumonidae), a parasite of the alfalfa weevil (Coleoptera:Curculionidae). Great Lakes Entomol. 9:189-93.
11. Koehler, C.S. and S.S. Rosenthal. 1975. Economic injury levels of the Egyptian alfalfa or the alfalfa weevil. J. Econ. Entomol. 68:71-75.
12. Hintz, T.R., M.C. Wilson, E.J. Armbrust. 1976. Impact of alfalfa weevil larvae feeding on the quality and yield of first cutting alfalfa. J. Econ. Entomol. 69:749-54.
13. Liu, B.W.Y. and G.W. Fick. 1975. Yield and quality losses due to alfalfa weevil. Agron. J. 67:828-32.
14. Fick, G.W. and B.W.Y. Liu. 1976. Alfalfa weevil effects on root reserves, developmental rate, and canopy structure of alfalfa. Agron. J. 68:595-99.

DISCUSSION

UNIDENTIFIED: First, does integrated pest management save the farmer money in alfalfa? And, second, what will the development of parasites mean to the farmer?
E. ARMBRUST: In the old programs, recommendations were to spray when anywhere from 10% to 60% of the tips showed feeding damage. The present program will allow much more effective use of pesticides, and generally eliminates one or two of the usual two or three sprays because of better timing. Concerning parasites, a number of them are established in various areas, but no one is totally effective. There are areas with five or six parasites established where the weevil is no longer as severe as it once was, and growers may be getting by with one less insecticide treatment.
UNIDENTIFIED: Are cost/benefit studies incorporated into this program?
E. ARMBRUST: Not really. Purdue has made a considerable effort in this area. One thing they have found is that when growers have to pay for a program, they follow it more closely.
L. BOWEN: Concerning the parasite, *Bathyplectes*, is there a difference in its effectiveness on the Egyptian and the eastern alfalfa weevil?
E. ARMBRUST: The two species are part of a complex; some entomologists say they are the same species...

C. HUFFAKER: The parasite is not effective against the
Egyptian alfalfa weevil. Even if the taxonomists can't tell
the difference morphologically, apparently the parasite can
distinguish between the two species.
E. ARMBRUST: The explanation may be that climatic limitations
are keeping the parasite from becoming established in some
areas.
L. BOWEN: In California, alfalfa is a key crop in parasite-
predator relationships. If you treat this crop early, you
might not have a reservoir of parasites left for other crops.
E. ARMBRUST: In these areas, although not in Illinois, the
crop does act as a nursery for parasites. In California,
some models have included consideration of these natural
enemies.
J. PORTER: The alfalfa growers here in New York have more of
the problem with leafhoppers than with weevils. Is there any
chance of a program on this?
E. ARMBRUST: Yes. This is a problem we are working on right
now, and it is also being worked on at Purdue University.
Alfalfa is now a cash crop, valued at $80 per acre, up from
only $15-20, so a lot of other insect problems are being
looked at now.
R. TROETSCHLER: Are certain insecticides less damaging to
natural enemies than others?
E. ARMBRUST: There are five or six insecticides recommended
for alfalfa weevil. The data are sketchy, but Imidan is
probably least disruptive to *Bathyplectes curculionis*.

POTENTIALS FOR RESEARCH AND IMPLEMENTATION OF
INTEGRATED PEST MANAGEMENT ON DECIDUOUS TREE-FRUITS[1]

Brian A. Croft[2]

Pesticide Research Center and
Department of Entomology
Michigan State University
East Lansing, Michigan

Alternative strategies of pest control and their use in integrated pest management (IPM) systems for deciduous tree fruits will be important in the future. Before discussing these alternatives, however, I will first summarize the biological characteristics of fruit agroecosystems and the pest control practices now common in apple orchards.

Apples are grown as a perennial tree crop, which is relatively stable from season to season. The trees are pruned, fertilized, and may be weeded annually, but they are not cultivated, nor are the plants replaced annually.

In the north central and eastern portions of the United States, apples harbor 40-50 potential arthropod pests, 10-20 potential disease pests, and a larger number of potential weed pests (1). Nematodes, birds, and mammals also attack apple trees.

Fruit tree pests may be classified according to three criteria (Table 1): (1) the severity and frequency of pest attack; (2) whether the pests attack the fruit or other parts of the trees; (3) the potential for control of the pests by biological means (predators, parasites, or disease). Annual key pests attack the fruit directly, nearly every year,

[1]*Published as Journal Article No. 8161 of the Michigan State University Agricultural Experiment Station.*
[2]*Currently on leave as Research Director of the Benson Agriculture and Food Institute, Brigham Young University, Provo, Utah 84602.*

101

Table 1.   Principal Disease and Arthropod Pests Occurring in
Apple Orchards of Midwestern and Eastern North America

Annual Key Pests

    Diseases
        Venturia inaequalis, apple scab
        Podosphaera leucotricha, powdery mildew

    Insects
        Laspeyresia pomonella, codling moth
        Argyrotaenia velutinana, red-banded leaf roller
        Conotrachelus nenuphar, plum curculio
        Rhagoletis pomonella, apple maggot

Potential Key Pests

    Large complex of species; see state or Canadian provin-
    cial fruit bulletins for individual area species
    complexes

Secondary Induced Arthropods

    Scales
        Quadraspidiotus perniciosus, San Jose scale
        Lepidosaphes ulmi, oystershell scale

    Aphids
        Aphis pomi, apple aphid
        Anuraphis roseus, rosy apple aphid
        Eriosoma lanigerum, woolly apple aphid

    Leafhoppers
        Typhlocyba pomaria, white apple leafhopper
        Empoasca fabae, potato leafhopper

    Mites
        Panonychus ulmi, European red mite
        Tetranychus urticae, two-spotted spider mite
        Aculus schlechtendali, apple rust mite

---

and have limited potential for control by biological means.
Arthropods in the second category, potential key pests, are
more sporadic in their occurrence; these pests cause either
direct or indirect damage, and may or may not have effective

biological agents. The third category consists of secondary induced pests. These arthropods generally have effective natural enemies controlling them in unsprayed orchards, but when broad-spectrum pesticides are applied for control of key pests, these natural control agents are eliminated or rendered ineffective. Thus, frequently secondary pests must also be controlled chemically.

On a per-acre basis, apples are one of the most heavily sprayed crops in the United States. They are a luxury commodity, valued for their dessert quality and cosmetic appeal. Under present marketing standards, fewer than 1% of fresh-market apples may be infested or damaged by pests. To maintain such a high standard of pest control, growers use preventive chemical pesticide programs. Thus it is not unusual for 20-30 different spray treatments to be applied to apple orchards each season.

A number of environmental issues are of concern when apple orchards are treated with pesticides. Orchards are often grown in the same location for long periods of time, and therefore often accumulate pesticide residues in the soil. In addition, because of their relative permanence, orchards are a preferred habitat for many nesting birds and other wildlife. Orchards are often grown near large bodies of water, such as the Great Lakes in Michigan and New York and the Columbia River in central Washington because of favorable climate and water drainage conditions in such areas. This geographic coincidence greatly magnifies the impact of orchard pesticides on associated aquatic systems.

The following section will discuss historical features of apple pest control with respect to four basic areas: pesticide use, basic biological research, IPM research, and IPM implementation. Also discussed are resistance of pests to pesticides, alternative methods of apple pest control, and the integration of these new methods into systems of IPM for delivery to growers. Finally, the lag between the development of IPM research and its implementation will be discussed. Some recommendations will be made for increasing the adoption of IPM on tree-fruit crops.

HISTORICAL TRENDS IN APPLE PEST CONTROL

In figure 1, trends in the four basic areas cited above are plotted for the period 1900 to the present. Each curve is my best estimate of the general trend, as determined from the literature. Pesticide use has followed a similar pattern for most crops. Insecticide (mostly chlorinated insecticides) use increased greatly during the post-World War II period

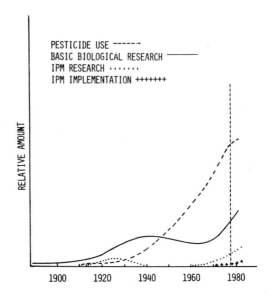

Fig. 1. Historical trends of four parameters associated with control of deciduous tree-fruit pests.

(1944-54). A switch to organophosphate compounds occurred in the late 1950s and 1960s because of resistance to DDT in certain pests [e.g., codling moth, red-banded leafroller, (2)], but total pesticide use continued to increase. In the 1970s, new compounds have continued to proliferate, with the development of pesticides derived from a variety of chemical bases, (e.g., carbamates, organotins, etc.). In the last 3-5 years, the total amount of pesticides used on apples may be leveling off (see also below, discussion on resistance). With fungicides, the spectrum of compounds has not been as wide, although the total use curve has the same form as the insecticide curve. Herbicides are still in a period of growth with regard to the appearance of new products. Relatively small amounts of herbicides are applied to apples compared with insecticides and fungicides.

Development of increasing numbers of new insecticides and increases in the amounts of insecticide used for fruit pest control are closely correlated with the development of pest resistance to the chemicals. This generalization holds for many other crops as well as tree fruits. In figure 2, the number of cases of resistance in insect, disease, and weed pests of all crops is graphed over time. The development of resistance among insects began with the San Jose scale on

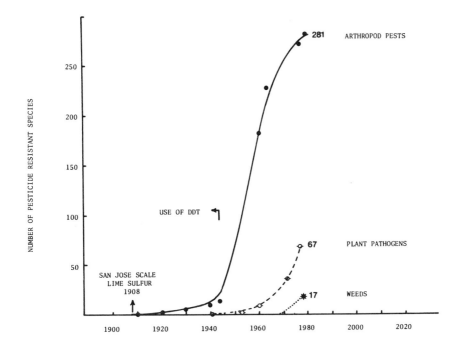

Fig. 2. The development of pesticide resistant species of arthropod, pathogen, and weed pests since 1908 [taken from (4) and (10)].

apples in 1908, and increased rapidly during the 1950s and 1960s, when organophosphate insecticides were widely used. Mites, leafhoppers, and aphids were notorious for the rate at which they acquired immunity during this period. In recent years, the rate at which resistance has been developing among apple pests appears to have declined in the United States, probably because integrated control systems are being used for species such as mites and aphids (3). The first appearance of resistance in disease organisms was later, and the number of cases has been increasing at a slower rate than for insects; there is evidence, however, that disease pests are entering a period of increasing resistance (4). For weeds, there are still only a few cases of tolerant species (4).

The extent to which basic biological research has been done on arthropod pests has been inversely related to pesticide use (fig. 1). Before our dependence on synthetic insecticides, considerable basic research was done on the biology and ecology of deciduous fruit species. For example, a

codling moth national research project was sponsored by the
Bureau of Entomology (USDA) from 1909 to 1929. In this program, about 50 research groups in the United States developed
basic studies on a countrywide basis. Some excellent biological research was done, and much of this information is
useful today in constructing population models (5,6). In the
1940s, biological research was largely replaced by chemical
screening and evaluation work. Before pesticides were widely
used, a detailed understanding of pest biology and ecology was
necessary to obtain even marginal pest control. With pesticides, however, preventive control with repeated applications
allowed considerable indifference to the occurrence and density status of pests. In effect, we simply replaced the need
for detailed biological information with a blanket of pesticides.

        With increasing pest resistance, the recognition of
environmental problems, and some pest control failures (e.g.,
mites, pear psylla), integrated pest control research on
deciduous fruit pests began on a broad scale in the early
1960s (fig. 1). An exception was in Nova Scotia, where unique
circumstances forced Canadian researchers to begin such
efforts in the 1930s and 1940s (7). More recently, the greatest success in finding alternatives to unilateral use of pesticides has come by integrating biological agents with chemical measures for control of indirect pests such as mites and
aphids (8,9). In the 1970s, research on IPM has increased
substantially, as has the basic research necessary to complement applied studies (fig. 1).

        Implementation of IPM is in its infancy, and has been
fraught with difficulties for the last 5-10 years (fig. 1).
Some progress in biological monitoring or scouting of fruit
pests has been made. Perhaps the most serious problem in
implementing IPM has been its requirements for technical
expertise, information processing, and education, which are
much greater than for chemical control procedures. Not only
must a grower or a pest manager using IPM know more precisely
what the pests are doing, he must also know which beneficial
species are present, and how to manage these species by modi-
fying pesticide applications.

        Recommendations on how to improve IPM are deferred to a
later section, on the future of IPM on tree fruits. I will
now review some new types of strategies and tactics available
for fruit pest control, and discuss the status of the development of IPM systems for the entire crop.

COMPONENTS OF AN IPM SYSTEM

The basic units of an IPM system from a systems analysis perspective (11) are: (1) a conceptual or mathematical model, including one or several pest control tactics and a knowledge of how these tactics affect the pest system under a variety of environmental and crop conditions; (2) a biological monitoring or pest survey system for determining the state of a given pest/crop system through time; (3) an environmental monitoring component that measures environmental conditions, allowing one to assess their influence on all components of the pest control system (pest, crop, natural enemy, control measure, etc.); and (4) an implementation system to apply the control strategy of the target pest species. Ultimately, we seek a control system for each pest in the crop ecosystem, integrated into a common plan. I will now review some prototype apple pest systems and a multiple species phenology system from my own research program in Michigan that demonstrates these elements. It should be pointed out, that although it is relatively easy to develop an alternative strategy for controlling a single pest, it is much more difficult to integrate such a strategy into an IPM system where pesticide use is still a dominant element.

PLANT-FEEDING MITES

The information concerning the pest management model presented in this section has been summarized in a number of other publications (1,12-16). The model involves a predator-prey system using the predaceous phytoseiid mite, *Amblyseius fallacis* Garman, which attacks the spider mite pests, *Panonychus ulmi* (Koch) and *Tetranychus urticae* Koch, along with chemical control measures to control most of the other apple fruit pests listed in Table 1. This integrated program is facilitated by the fact that *A. fallacis* is resistant to a variety of organophosphate insecticides (3,17). The pest management system components (see previous section) for this pest system have also been described elsewhere. They include optimal sampling programs for predators and pests, providing minimum sample sizes for known levels of precision. The system also incorporates features such as monitoring unit size, maximum time delays, relative sampling intensity, and economic risk (18,19). An on-line environmental monitoring network from 41 stations throughout the Michigan fruit belt (11,16,20) allows us to run the model based on real-time weather conditions. Finally, a computer-based extension delivery system integrates biological and environmental

monitoring information with the mite management model in a
central computer, and provides telecommunications access for
users at remote terminals in extension agent offices throughout the state's fruit belt (20).

THE CODLING MOTH

The codling moth pest management system has also been
described in several previous publications (5,6,16). The
system is based on the capture of male moths at pheromone-
baited traps. These data are used to track emergence of the
adult moths over the season. The data are coupled with a
physiological time model for insect development, driven by on-
line weather information. Information on development of the
moth and timing of pesticide applications to control hatching
larvae is conveyed to field users at regional extension
offices (20).

A MULTIPLE-SPECIES EXTENSION TIMING SYSTEM FOR APPLE IPM

This system, with the acronym PETE (Predictive Extension
Timing Estimator), is a phenologically based modeling system
for the entire apple pest complex (21). It was designed to
enable researchers to implement extension phenological models
with a minimum of background and effort. PETE programs implement a new model by inserting it into a Pest Management Executive System (PMEX) (20). The latter automatically obtains
the necessary biological and environmental inputs from field
sites, runs the model, and disseminates the resultant pest
alerts to the appropriate extension field staff in an on-line
mode.

There are presently 17 species programmed into PETE:
the tree, apple scab disease, and a number of arthropod pests
of the apple, including codling moth, oriental fruit moth,
apple maggot, and red-banded leafroller. Models for each of
these organisms are incorporated into the PETE system in four
stages: (1) literature search, (2) parameterization, (3) validation, and (4) implementation. At present, research for
each pest has progressed beyond step one, and seven species
models have been validated. One objective of this system is
to summarize our current understanding of the phenology of
fruit pests in Michigan. Another is to predict more precisely
the temporal activities of these species. Although our precision in forecasting pest development and identifying important management "windows" will be limited initially, we believe that this system will represent a significant

improvement over current spray manuals, which are based on tree blossom development. This effort will also provide a valuable tool to identify deficiencies in our understanding of fruit pest phenology, and thus to direct future research.

OTHER ALTERNATIVE STRATEGIES FOR APPLE PEST CONTROL

Beyond increased use of biological controls and improved timing systems for IPM, which are under development in most states with major planting in fruit trees [reviewed in (22)], several experimental control methods are on the horizon. The sterile-male technique for the codling moth is technically feasible, but is not economically competitive with traditional chemical control methods (23). Pheromones are widely used for improved timing of pest management tactics. Control systems using the confusion technique with pheromones have been demonstrated experimentally for certain Lepidopteran species (24). Pheromone control systems have not been developed for an entire pest complex, however, and unless they are, it is unlikely that they will supplant pesticides. Host-plant resistance is currently more useful against disease pests of apple than against arthropods. Researchers are just beginning to study the possible exploitation of biological control measures for disease pests (25). For weed pests of orchards, allelopathy and biological control by arthropods that feed on weeds are two areas that may offer some possibilities (25).

INTEGRATION OF APPLE IPM SYSTEMS

At present, we are still working with bits and pieces of a truly integrated pest management system for an entire fruit pest complex. As noted earlier, we now have pest management models, biological monitoring systems, environmental monitoring systems, and extension delivery methods for several indirect pests, such as plant-feeding mites, and for certain key pest species, such as codling moth and apple scab. We are still working on the basic biology of the remaining majority of pests, and have adequate sampling methods for only a few. Fortunately, the needs for environmental monitoring and extension delivery are the same for most pests, and we have made considerable progress in these areas.

In addition to the basic components of a pest management system, we must consider the economic and optimization phases to determine the best combination of methods to use for given pest conditions. These tools must be applied to a pest control system in the final step before it is implemented to obtain the proper combination of control techniques. We are beginning to be more rigorous in evaluating the economics of individual control strategies for apple pests, but we have not yet begun to approach optimal combinations of pest control measures in a fully integrated system (26).

FUTURE TRENDS AND RECOMMENDATIONS

Now that some of the newer innovations in apple IPM have been reviewed, I turn to a discussion of the future. What are the goals we should set for this area of applied science, and what steps must be taken to reach them?

To answer the first question, I have extrapolated figure 1 into the future (fig. 3). The vertical hatched lines indicate the next 15-year period, our short-range planning horizon. The pesticide use curve begins to decline in that period. Beyond 2000 we expect to attain some realistic, lower level of pesticide use. Some pesticide use will probably always be necessary on tree-fruits unless economic thresholds change in the future. Of course, many improvements may be made in the chemicals that are used for apple IPM. Continued effectiveness, lower mammalian toxicity, reduced environmental side effects, and limited ecosystem persistence should be emphasized. It is anticipated that applied IPM research will continue to increase at a rapid rate, and that basic research on pests and associated species such as natural enemies will be commensurate with this effort.

Probably the biggest challenge facing those working with IPM will be to reduce the lag time between research and implementation so that progress in these two areas can occur more in parallel (see fig. 3). One major problem relates to the training and education of personnel to provide technical IPM assistance in the field. If a grower cannot cope with the increased complexity of IPM methods, in some areas he can buy such expertise from professional consultants or applied insect ecologists. These persons currently are relatively scarce, because to operate such a service requires many years of experience in the field. Training such personnel by traditional education systems (colleges, trade schools, etc.) is difficult (fig. 4). In the future, educational methods must be developed to substitute for the vast experience presently required to become a competent IPM consultant.

It is my opinion that improved basic and IPM research, systems science techniques, and new education methods such as gaming offer hope for supplying experience in the field via simulation problems in the laboratory. We are still in the beginning stages of developing such tools at Michigan State

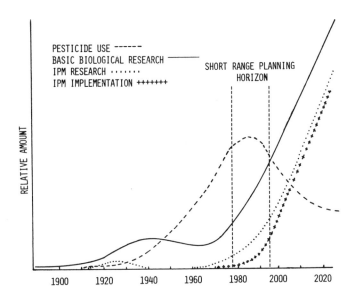

Fig. 3. Historical and anticipated future trends of four parameters associated with control of deciduous tree fruit pests.

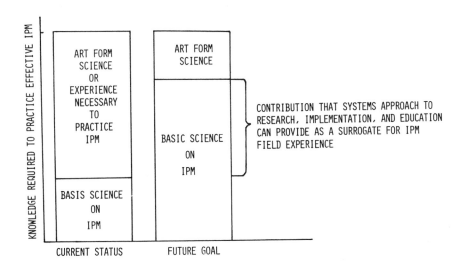

Fig 4. Graphic representation of relative proportions of basic science and experience necessary to practice IPM on tree fruits presently as compared with the desired proportion to be established in the future.

University, however, we have developed three programs for training purposes. These include a cereal leaf beetle popu- lation simulation game ("Popdyn"), an insecticide resistance gaming program ("Resistance"), and a predator-prey inter- action program ("Orchard"). Obviously some field experience will always be needed in training pest managers, but this kind of educational tool can often provide comparable experi- ence, and can expose students to circumstances they would wait many years to see in nature. Such tools are only part of the improved education we must provide. Extension educa- tion also has the tremendous task of upgrading the competency of farmers and other technical persons who are now operating in the field.

To accomplish the goals of greater research and imple- mentation of IPM on tree fruits, I propose the following more specific recommendations for the future. This is by no means an all-inclusive list. In essence, it is a summary review of the points made earlier:

1. We need continued and even increased support for basic biological research dealing with the system- atics, biology, toxicology, behavior, and ecology of agricultural pests.

2. We need support to develop alternative tactics of fruit pest control. The technology of IPM also needs further development, to provide the means to inte- grate these new methods into simplified systems for practical pest control.

3. We need continued and even accelerated development of "selective pesticides." By selective, I mean as selective as possible, given the economics of pesti- cide development. The possibility of subsidized developmental costs still needs to be considered by federal agencies.

4. We need to develop additional ways to structure IPM research and increase our ability to develop predic- tive systems of IPM. Much progress in this area will come through the application of systems science methodologies.

5. We need well-trained biomonitoring personnel or scouts to carry out detailed observational surveys in the field. This will provide many jobs at the farm level.

6. We need improved and expanded agricultural weather monitoring facilities, and companion systems for summarizing and delivering these inputs for use by pest managers.

7. We need to include better economic assessment of pest control strategies and especially to optimize the

combined use of IPM tactics, their timing, and level of application.

8. We need improved extension delivery systems that are efficient, fast, and capable of handling large information sets for IPM.

9. We need improved educational programs to teach growers and IPM practitioners how to use the new management tools to greatest effect.

REFERENCES

1. Croft, B.A. 1975. Tree fruit pest management. pp. 471-507 *in* Introduction to Pest Management. R.L. Metcalf and W.H. Luckmann, eds. Wiley Interscience, New York.

2. Hough, W.S. 1963. Resistance to insecticides by codling moth and red-banded leafroller. Va. Agr., Exp. Sta. Tech. Bull. 166.

3. Croft, B.A. and A.W.A. Brown. 1975. Responses of arthropod natural enemies to insecticides. Annu. Rev. Entomol. 20:285-335.

4. Anonymous. 1977. Pest resistance to pesticides and crop loss assessment—1. Report of 1st FAO Panel of Experts. FAO Plant Production and Protection Paper 6.

5. Riedl, H. and B.A. Croft. 1977. Management of the codling moth in Michigan. Mich. State Univ. Agr. Exp. Sta. Tech. Bull. (in press).

6. Riedl, H., B.A. Croft, A.J. Howitt. 1976. Forecasting codling moth phenology based on pheromone trap catches and physiological time models. Can. Entomol. 108:449-60.

7. Pickett, A.D., W.L. Putman, E.S. LeRoux. 1958. Progress in harmonizing biological and chemical control of orchard pests in eastern Canada. Proc. Int. Congr. Entomol. 3:169-74.

8. Asquith, D. 1971. The Pennsylvania integrated control program for apple pests—1971. Pa. Fruit News 50:43-47.

9. Hoyt, S.C. 1969. Integrated chemical control of insects and biological control of mites on apple in Washington. J. Econ. Entomol. 62:74-86.

10. Brown, A.W.A. 1977. How have entomologists dealt with resistance. Proc. 68th Ann. Meet. Am. Phytopathol. Soc. 3:67-74.

11. Haynes, D.L., R.K. Brandenburg, D.P. Fisher. 1973. Environmental monitoring network for pest management systems. Environ. Entomol. 2:889-99.

12. Croft, B.A. 1975. Integrated control of apple mites. Mich. State Univ. Coop. Ext. Serv., Ext. Bull. E-825.

13. ———. 1975. Integrated control of orchard pests in the

USA. Proc. 5th Symposium on Integrated Control in Orchards
5:109-24.

14. ————. 1976. Pest management systems for phytophagous
mites and the codling moth. pp. 99-121 *in* Proc. 3rd
US/USSR Pest Management Conference on the Integrated
Control of the Insect Pests of Cotton, Sorghum and Deciduous Fruit. Tex. Agr. Exp. Sta. Ser., MP-1276.

15. Croft, B.A. and D.L. McGroarty. 1977. The role of
Amblyseius fallacis in Michigan apple orchards. Mich. Agr.
Exp. Sta. Bull. (in press).

16. Croft, B.A., R.L. Tummala, H. Fiedl, S.M. Welch. 1976.
Modeling and management of two prototype apple pest
subsystems. pp. 97-119 *in* Modeling for Pest Management.
R.L. Tummala, D.L. Haynes, B.A. Croft, eds. Michigan State
University, East Lansing. Proc. 2nd USSR/USA Symposium on
Integrated Pest Management.

17. Croft, B.A., A.W.A. Brown, S.A. Hoying. 1976. Organophosphorus resistance and its inheritance in the predaceous
mite *Amblyseius fallacis* on apple. Environ. Entomol.
69:64-68.

18. Croft, B.A., S.M. Welch, M.J. Dover. 1976. Dispersion
statistics and sample size estimates for population of the
mite species *Panonychus ulmi* and *Amblyseius fallacis* on
apple. Environ. Entomol. 5:227-34.

19. Welch, S.M. 1977. The design of biological monitoring
systems for pest management. Ph.D. Dissertation, Michigan
State University.

20. Croft, B.A., J.L. Howes, S.M. Welch. 1976. A computer-
based extension pest management delivery system. Environ.
Entomol. 5:20-34.

21. Welch, S.M., B.A. Croft, J.F. Brunner, M.F. Michels. 1978.
Management of multi-species pest complexes via a generalized phenology modeling system. Environ. Entomol. (in
press).

22. Asquith, D., B.A. Croft, S.C. Hoyt, E.H. Glass, R.E. Rice.
1976. The principles, strategies and tactics of pest population and control in the stone and pome fruits subproject. C.B. Huffaker, ed. University of California, Center
for Biological Control.

23. Hoyt, S.C. and E.C. Burts. 1974. Integrated control of
fruit pests. Annu. Rev. Entomol. 19:231-52.

24. Trammel, K. 1972. The integrated approach to apple pest
management and what we are doing in New York. Proc. N.Y.S.
Hort. Soc. 117:37-49.

25. Anonymous. 1977. Biological agents for pest control:
status, feasibility and implementation. USDA Tech. Rept.
(in press).

26. Croft, B.A. 1977. The economics of integrated pest

management in orchards *in* Proc. 4th USA/USSR Symposium
on Integrated Pest Management. Yalta, USSR, in Russian
(in press).

DISCUSSION

R. WHETSTONE: Were there changes in fruit yield and quality
in the integrated mite control program?
B. CROFT: We actually found that integrated mite control left
fewer eggs in the calyx end of the fruit, where they affect
fruit quality. With chemical control, the surviving mites
reach a population peak at the end of the season, which is
the only time they can affect the apples. With biological
control, the mites peak in mid-season, and you have predators
"mopping up" at the end of the season.
B. DAY: Do you have any data on the effects of mites on yield?
B. CROFT: It is difficult to measure the effects of mites.
During one season they can affect color and, at high density,
yield of fruit, but effects on the trees are also cumulative.
So many other factors affect the trees, the question is really
almost unanswerable. We did check how often the mites exceeded
the economic injury level, and found that this occurred more
often with chemical programs than with integrated pest control
programs.
S. BEER: What role, if any, do you see for cultivar selection
in apple pest management?
B. CROFT: There are many pests for which resistant cultivars
exist, and there are many projects going on, concerning both
insect and disease resistance. In the past, these have not
been integrated well into fruit pest management, but they will
be in the future. They are very long-term projects, but I
feel they should be pursued more.

POTENTIALITIES FOR PEST MANAGEMENT IN POTATOES

H. David Thurston

Department of Plant Pathology
Cornell University
Ithaca, New York

In this paper I plan to show that we have rather sophisticated programs for management and control of some potato
pests, especially diseases. Having been a member of The Potato Association of America for 27 years, my first action in
preparing this talk was to go through the index of the *American Potato Journal* from 1967 to the present, looking for any
mention of the words pest management or integrated control in
the title of any major article. There was none. I should
hasten to add that these terms are used in relation to potatoes. Finding no mention of integrated pest management
(IPM) in the major journal used by almost all potato researchers does illustrate that the language or terminology of IPM is
not yet commonly used by potato scientists, extension workers,
and least of all by farmers. Even though the terminology of
pest management is seldom used, many of the principles behind
the concept have long been employed by the potato industry for
controlling pests, and programs of potato pest management are
evolving.

As a plant pathologist, my emphasis will be on management
of plant pathogens, but management of other pests will also be
discussed.

Why should we be concerned with IPM in potatoes? Large
quantities of pesticides—insecticides, fungicides, nemati-
cides, and herbicides—are used on potatoes today, and there
exists a real possibility of significantly reducing the quantity used. For example, 2,889,000 pounds of insecticides,
4,124,000 pounds of fungicides, and 6,285,000 pounds of herbicides were used on potatoes in 1971 in the United States (1).
Nematicide use is also significant. Only peanuts, citrus, and
apples use more fungicides than potatoes in the United States.
According to Andrilenas (2) 96% of the potato acreage in the

northeastern United States is sprayed with protectant fungicide.

Reduced pesticide use will become more important with increasingly resistant pest strains, rising costs of pesticides and energy to the farmer, and with public demands for environmental and health protection.

The white or Irish potato *(Solanum tuberosum)* is one of our major food crops. In terms of total annual tonnage produced, the potato ranks fourth in the world after wheat, rice, and maize (3). The potato is a relative newcomer among major food crops as it was restricted to the Andean highlands until introduced into Europe by the Spanish in the sixteenth century.

When the potato was introduced into Europe from South America, some of its pests came with it, such as black wart *(Synchytrium endobioticum)*, the potato cyst nematode *(Globodera pallida—G. rostochiensis)*, and perhaps certain tuberborne viruses. However, many pests were left behind in the Andes, such as rusts *(Puccinia pittieriana, Aecidium cantensis)*, smut *(Thecaphora solani)*, the Andean weevil complex, and false root-knot nematodes *(Nacobbus spp.)*, which are thought to cause as severe yield reductions as the potato cyst nematode. As the potato spread into Europe and North America, it was attacked by new pests not found in the South American Andes such as *Phytophthora infestans* (late blight), the Colorado potato beetle *(Leptinotarsa decemlineata)*, and *Corynebacterium sepedonicum* (ring rot).

Pest control with potatoes is complicated and difficult, as the potato is vegetatively reproduced, using tubers for seed. Bacteria, viruses, fungi, insects, and nematodes are easily carried with seed tubers, and some are rapidly disseminated by cutting knives and picker-planters. Sources of relatively pest-free seed produced by seed certification programs are essential for economic production of healthy potatoes. From 2500 to 3000 pounds of tubers are used to plant an acre of potatoes, whereas only 120 to 180 pounds of seed per acre are used with cereals such as wheat, oats, or barley. The sheer bulk of the seed of a vegetatively reproduced crop, which is difficult to store more than 6 months, makes seed production programs with potatoes far more difficult than those with cereal or other crops planted with true botanical seed.

Major components of pest management programs are the use of cultural practices or agroecosystem manipulations, host resistance or genetic control, biological control, legal regulations, and the use of pesticides. Potato pests are presently controlled by all of these means.

RESISTANT POTATO CULTIVARS

Of the component parts of an IPM program, varieties with resistance to pathogens are a major element. Unfortunately, to date it has been impossible to incorporate needed resistance to all pathogens into a single potato cultivar. Breeding new potato varieties is expensive. R.L. Plaisted[1] has estimated that it normally takes 200,000 seedlings to find one worthy of release and 1 million seedlings to find a new variety that will eventually be used on a significant percentage of U.S. potato acreage. It generally takes 15 years or more to progress from the time a cross is made until it produces a named variety. The cost of producing and evaluating 200,000 seedlings has been estimated recently for New York State at $200,000[1].

Incorporation of multiple pest resistance into new potato varieties is highly desirable, but breeders must also incorporate a long list of other characters into new varieties. High yielding ability, wide adaptation, cooking quality and flavor, processing quality, good storage characteristics, tuber size, pleasing shape, skin color, flesh color, depth of eyes, freedom from external and internal defects, acceptable vine type, freedom from toxic chemicals (i.e., glycoalkaloids), plus other agronomic characters may be more important to potato producers and consumers than pest resistance. With over 30 viruses and 80 bacteria and fungi attacking potatoes in the United States (4) it is obviously impossible to incorporate resistance to all and probably even to the most important pathogens. When nematodes and insects are also considered, the task becomes even more difficult and setting priorities becomes essential.

In the United States, there are now 12 major potato breeding programs, and from 1967-1976 26 new potato varieties were released from these and other private breeding programs. Akeley (5) reported on the development of potato varieties in the United States up to 1966 with multiple-disease resistance. Of 18 varieties released from 1932-1966, resistance to 14 diseases was reported, and all varieties had resistance to at least two and up to six diseases.

The 26 varieties produced since 1967 incorporate differing levels of resistance to 16 plant pathogens, two species of nematodes, and a degree of resistance to two insect species, according to the *American Potato Journal* variety descriptions. Ten of the 26 varieties have multiple disease resistance with resistance to four or more diseases. These figures should not lead one to conclude that varieties with high levels of

[1]*personal communication*

resistance to all major potato diseases are available, for this is not so, and in fact only one of these 26 varieties consti- tutes more than 5% of the total acreage of certified seed pro- duced in the United States and Canada (6). Nevertheless, many varieties with resistance to scab (*Streptomyces scabies*), net necrosis (caused by potato leafroll virus), mild mosaic (Po- tato virus A), verticillium wilt *(Verticillium dahliae* and *Verticillium albo-atrum)*, latent virus (virus X), and the po- tato cyst nematode are available and often play a significant role in the control of these pests. One variety (7) with resistance to hopperburn caused by the leafhopper *(Empoasca fabae)* and two varieties (8,9) with tolerance to the potato flea beetle *(Epitrix cucumeris)* are currently available.

Many of these varieties can play an important role in a program of IPM. With more adequate funding for research, how- ever, the potential exists to produce varieties with multiple resistance to many more pests, thus effecting a significant reduction in pesticide use.

The potato varieties now grown in North America and Europe are highly selected cultivars originating from a narrow genetic base. Their ancestors, the introductions brought to Europe by the Spanish, probably did not represent a broad sampling of the range of germplasm of potatoes cultivated in the Andes of South America at that time. This narrow genetic base has been hybridized, selected, and reselected for adapta- tion and desirable characters, thus reducing the genetic base even further. It has been estimated that only 5% (10) of the available potato (*Solanum* spp.) germplasm has been utilized in today's potato varieties.

The first potato introductions to North America came to New Hampshire in 1719 from England. It was not until the 1850s that the Rev. Chauncey Goodrich, of Utica, New York, obtained a few tubers from South America and introduced ad- ditional germplasm to widen the genetic base of U.S. potatoes (11). One of the seedlings from Goodrich's material, Garnet Chili, is a parent of many of the major U.S. varieties today (12).

Solanum tuberosum subsp. *andigena* (andigina), the potato species most commonly grown in the Andes of South America and the species from which *S. tuberosum* subsp. *tuberosum* (tubero- sum) of Europe and North America probably evolved, was seldom utilized by potato breeders until the 1960s, when Simmonds (13), at the John Innes Institute in England, was successful in adapting andigena to tuberize under long-day conditions.

The resistance to the potato cyst nematode first used in Europe and the United States came from andigena, and breeders noted that andigena derivatives yielded somewhat more than other breeding material. In 1963 a program was started at

Cornell with material from the Commonwealth Potato Collection plus material collected in Peru, Ecuador, and Colombia to utilize andigena in breeding to widen the germplasm base in the Cornell potato-breeding program. Recurrent selection has improved the yield and appearance of this andigena population. Screening seedlings for late blight resistance in the greenhouse at Ithaca and evaluating selected clones under epiphytotic conditions in Mexico, where all known races of the pathogen are found, have developed valuable levels of general resistance. Tests for resistance to verticillium wilt and to scab have identified new sources of resistance to these diseases. An unplanned epidemic of virus Y revealed several andigena clones with resistance. Virus X and potato wart resistance have been identified. Nematologists have found resistance to the root knot nematode and entomologists have identified resistance to leafhoppers, aphids, and the tarnished plant bug (12,14).

This new source of potato germplasm opens up an entirely new source of resistance to pests. The goal of multiple pest resistant potatoes is not unobtainable.

Probably the single most important disease of potatoes in the United States (and worldwide) is late blight, caused by *Phytophthora infestans*. The Irish "potato famine" of the 1840s was caused by *P. infestans*. Although in most of the world late blight is controlled by fungicides and not by resistant varieties, efforts are continuing to develop resistance.

Two types of resistance to *P. infestans* in potatoes are recognized. These are specific resistance (also called race specific, vertical, oligogenic, or monogenic resistance) and general resistance (also called race nonspecific, horizontal, or polygenic resistance) (15). Before the discovery of specific resistance in the wild species *Solanum demissum,* which could be incorporated into *S. tuberosum*, general resistance was the only type of resistance available, and fairly high levels were observed. For several decades after the discovery of specific resistance in *S. demissum,* breeders incorporated one or a few *S. demissum* genes into new potato varieties. *P. infestans* is a highly variable organism; thus the use of specific resistance contributed little to controlling late blight, because the pathogen rapidly overcame such resistance. The hopes of obtaining lasting specific resistance are very slim, as all potato cultivars and all tuber-bearing *Solanum* species have been found to be susceptible to late blight in the Toluca Valley of Mexico, where the sexual stage of *P. infestans* is found (16). There are currently no cultivars in Europe or North America that allow commercial cultivation of potatoes without fungicide use where late blight is a problem.

Some commercial varieties such as Sebago have a moderate level of general resistance, and less fungicide can be applied to them, as has been documented in New York by Fry (17) and on Prince Edward Island, Canada (18). Breeding efforts in several continents are being directed toward obtaining cultivars with high levels of generalized resistance, which can be used with reduced amounts of fungicide or even without fungicides in drier areas.

Opportunities to incorporate resistance to potato pests into improved potato varieties exist. In addition to high levels of generalized resistance to *P. infestans,* resistance is available that could be used against diseases and nematodes such as scab and verticillium wilt, early blight *(Alternaria solani),* silver scurf *(Helminthosporium solani),* viruses X, A, Y, and leafroll, southern bacterial wilt *(Pseudomonas solanacearum),* four species of root knot nematodes *(Meloidogyne* spp.), and meadow nematode *(Pratylenchus penetrans).* Extensive studies have led to identification of numerous sources of insect resistance (19-27), but little of this resistance is incorporated into the varieties grown today. Promising sources of resistance exist within wild, tuber-bearing and tetraploid *Solanum* germplasm to various species of aphids, leafhoppers, and other important potato insects.

CULTURAL PRACTICES

There is a wide variety of cultural practices and agro-ecosystem manipulations used to control potato pests. I cannot cover all of them in this paper, but rather will discuss a few to illustrate how they can be integrated into pest management programs.

Use of Certified Seed

Basic to all potato production is the availability of a reliable source of relatively disease-free seed (tubers). Probably the oldest and most extensive seed certification effort in crop protection history has been with potatoes, and its success has stimulated the formation of certification programs for many other crops (28). Seed certification programs were started in Europe in the early 1900s and in the United States about 1914. There are presently 21 state seed certification programs with various forms of administrative organization (29). The agencies that certify seed set minimum standards for specific diseases, abnormalities, and mixtures, and although standards are similar, they are not always the same in different programs. Nematodes are considered only in seven programs. All programs have a zero tolerance for the occurrence of ring rot (*Corynebacterium sepedonicum*), a highly infectious bacterial disease.

Certified seed producers obtain their seed from "foundation" seed growers who produce elite seed in isolated areas. These growers maintain extremely high standards for freedom from disease. All seed lots entered as foundation stock have a "winter test," in which tuber stocks to be entered as foundation seed are planted in the greenhouse or in the field in Florida or California and evaluated for disease incidence before the foundation seed is sold. Certified seed growers use foundation seed for their plantings, and inspectors from certification agencies generally inspect their fields two to three times a season. Growers can rogue out diseased plants between inspections, except for diseases with a zero tolerance, such as ring rot. There is usually an inspection after harvest for internal disorders and diseases. If tolerances are exceeded in the field or storage, the crop is not eligible for certification. Most diseases are diagnosed on the basis of symptoms in the field, but increasingly, rapid laboratory diagnostic techniques are being used, especially serology for viruses. The production of virus-free plants using meristematic tip culture and rapid propagation by vegetative cuttings is an important recent innovation in potato seed certification programs.

The production of certified seed is far more complex than it has been possible to indicate in this paper. Additional information is given by Shepard and Claflin (28) and Jones and Knutson (29). Seed certification is basic to the potato industry of the United States and is an essential component of any pest management program in potatoes.

Rotations

Rotations with potatoes are a component of both modern and traditional agriculture. They are not new.

In Peru, before the arrival of the Spanish, the Incas had a seven-year rotation for potatoes established by law. Through centuries of trial and error, the Inca Indians had discovered that this rotation gave the best potato crops. The potato cyst nematodes (*Globodera pallida—G. rostochiensis*) are present in extremely high levels in most potato-growing areas of the Peruvian Andes. With the arrival of the Spanish, Inca law was destroyed and the seven-year rotation was abandoned. It is now known that with a seven-year rotation the population of the golden nematode is reduced so that an economic crop can be produced[2]. Thus, what appeared to the Spanish to be a senseless custom had a sound practical basis and

[2]*W.F. Mai, personal communication*

was an intelligent agronomic and crop protection practice.

In a 1976 USDA publication on potato diseases (30), crop rotation is recommended as a means of disease control for 13 of the 69 diseases discussed. It is especially important for the long-term control of diseases such as verticillium wilt, sclerotinia white mold *(Sclerotinia sclerotiorum)*, common scab, powdery scab *(Spongospora subterranea)*, fusarium wilt *(Fusarium* spp.*)*, and the golden nematode.

Brodie (31) has studied the effects of various management systems on the population densities of *Globodera rostochiensis,* the golden nematode. These systems, which he followed for five years in replicated field plots, included a monoculture of a susceptible cultivar (Katahdin), a nonhost crop (oats), chemical soil treatments, and a resistant potato cultivar (Hudson). The results indicate that to manage *G. rostochiensis* successfully at densities below the spread level, a resistant cultivar must be grown at least once every other crop. To manage nematode populations below the plant damage level, susceptible varieties should not be grown in successive years without chemical (aldicarb) treatment.

The potato is grown under such a wide variety of soil and environmental conditions that rotation recommendations generally should be site specific. It is important that the crop rotation does not include plants that are also hosts of the potato pathogens, since that may make the problem more serious.

Manipulation of Soil pH

Common potato scab is caused by the actinomycete, *Streptomyces scabies*. The disease generally does not occur on soils with a pH below 5, but may be severe in soils with a pH of 6 or above. The use of sulfur to reduce pH can control scab, but is seldom economically feasible. The use of lime should be avoided on potato soils where its use would raise the pH above 5 to 5.2 (32).

Cultural Manipulations and Sanitation

The late blight disease *(P. infestans)* provides several examples of how cultural manipulations and sanitation procedures are used to reduce losses due to disease organisms.

In temperate or subtropical areas, where distinct climatic seasons occur, it is important to delay initial infection by *P. infestans* as long as possible. First, seed tubers free of *P. infestans* are essential. Second, inoculum sources such as infected tubers in cull piles near farm or commercial storages, which are a frequent source of primary inoculum in the spring, should be destroyed.

Tuber infection results from sporangia of the fungus being washed into the soil from blighted foliage. Consequently, good coverage of tubers with soil by adequate "hilling" is important to reduce tuber infection. Exceptionally large "hills" are commonly made in the Andes of South America, and as a result tuber infection by *P. infestans* is rare.

If foliage does become infected by *P. infestans* late in the season, tuber infection can be prevented or greatly reduced by killing the vines at least two weeks before harvest. This prevents further tuber infection; tubers already infected will rot sufficiently in the ground so that they will not be harvested. Before tubers are stored, they should be carefully examined and blighted tubers should be discarded.

Many other examples could be given of cultural manipulations and sanitation procedures of importance in controlling other potato pests. For example, the severity of *Rhizoctonia solani* can be decreased by shallow planting. Common scab severity can often be reduced by proper manipulation of irrigation (33). As potato tubers for seed are commonly cut into pieces for planting, the knives or machinery such as mechanical seed cutters used to cut seed can rapidly spread bacteria and viruses. Furthermore, most potatoes are planted using mechanical picker-planters with "picks", which are ideal for inoculating tubers with bacteria and other pathogens. These procedures make sanitation and the use of disease-free seed especially important.

Cultural methods such as manipulation of planting dates and hilling practices have been studied by Shands, Simpson and Murphy (34) in relation to aphid control, leaf roll virus spread, and severity of rhizoctonia disease. In Maine (35) and Idaho (36) the primary overwintering hosts (*Prunus* spp.) of the green peach aphid *(Myzus persicae)* are being eradicated systematically as a component of a pest management program.

Little has been said about the control of one of the major pests of potatoes: weeds. Almost as many pounds of herbicides are used on potatoes as fungicides and insecticides combined. In addition to herbicides, mechanical methods of plowing, harrowing, and cultivating are extensively used. Mechanical tillage costs are immense, especially with rising energy costs (37). Yip (38) and Sweet, Yip and Sieczka (39) have shown that potato varieties such as Hudson and Green Mountain, because of their large, vigorous vines, are competitive with several species of weeds when compared with varieties such as Norchip, Sebago and Katahdin. Such competitive cultivars may be useful in reducing herbicide or tillage costs.

BIOLOGICAL CONTROL

Potato researchers have little to report regarding practical control of potato pests using biological control. Considerable research has been done on the use of green manure
crops in rotation with potatoes for the control of common
scab, and a soybean cover crop appears to prevent a buildup of
S. scabies (40). Work in Maine has shown that aphid pests of
potatoes are controlled in nature by several species of pathogenic fungi (41), many species of parasites (42), and arthropod predators (43). Unfortunately, many insecticides reduce
the effectiveness of the parasites and predators. There is
little evidence that widespread commercial use is made of the
above findings regarding biological control.

LEGAL REGULATIONS

A major means of protection from pests is exclusion, or
preventing the introduction of a pest into a presently un-
colonized area. Many potentially serious potato pests such as
black wart, rusts, smut, and race 3 of *P. solanacearum* (bacterial wilt) have been excluded from the United States by
quarantines. Black wart was found in the United States in
1918 in home gardens in Pennsylvania, West Virginia, and Maryland. Eradication efforts and quarantines were successful,
and black wart has not occurred since then (32). The golden
nematode was first found on Long Island in 1941, but was confined to Long Island in New York State by quarantine measures
until 1967, when it was found in Steuben County in western
New York. Strong federal and state quarantine programs are
still in force, but the nematode has been reported in three
new counties of New York this year. Races of potato cyst
nematodes exist in Peru that can attack both European and U.S.
sources of genetic resistance; thus quarantines and legal restrictions against these pathogens are still important.
At the county and state level, foundation and certified
potato seed growing areas often maintain legal quarantines
against potato importations in order to protect seed certification programs. Legal regulations should be considered an
integral part of potato pest management.

CHEMICAL PESTICIDES

Pesticides are used on almost all the potatoes grown in
the United States. About 6,649 tons of chemical pesticides
were used on potatoes in the United States in 1971 (1).

Fungicides

As noted earlier, over 2,000 tons of fungicides were used in 1971 on potatoes in the United States. The majority is used to control late blight, but significant quantities are also used to control early blight *(Alternaria solani)*, and to treat tuber seed pieces.

Bactericides and various fumigants are used for sanitation purposes in potato warehouses and storage areas, on machinery and other equipment.

The major control method against late blight is the use of protectant fungicides. Bordeaux mixture was first tried as early as 1885 for late blight control, but its phytotoxicity and the discovery of better fungicides have subsequently made it obsolete. Fixed copper compounds and, later, organic compounds, especially the dithiocarbamates, have come into general use.

Recently, specific systemic fungicides have been tested experimentally to control *P. infestans* and, although results are promising, none of these chemicals are yet available for grower use. The use of systemic fungicides has advantages for growers because fewer applications are needed and simplified application equipment can be used. The use of systemic fungicides has dangers, however (44). The older, heavy metal fungicides and the organic fungicides have been used for many decades with only rare reports of fungi developing resistance (45). These older fungicides were essentially protectants, and for effective control a continuous film over the entire surface of the plant was necessary. In contrast, systemic fungicides (e.g., oxathiins, pyrimidines, benzimidazoles), currently used for the control of other fungi, penetrate the cuticle and are translocated throughout the plant, so that their action is much more efficient. These systemic fungicides were all introduced in the last 10-15 years and, because of their obvious advantages, rapidly became widely used on many crops other than potatoes.

The mode of action of the protective fungicides was generally nonspecific, interfering with many vital functions of fungi. In contrast, some systemic fungicides such as benomyl are highly specific in their mode of action. Thus, their fungicidal action seems to depend on the interference with only one or a very few vital functions (e.g., it appears that the benzimidazole compounds inhibit chromosome migration), and a single gene mutation in the pest organism can result in a modified system no longer sensitive to attack. Such a change would result in an immune individual and provide the basis of a resistant population.

It thus appears that selective fungicides will become

more widely used for control of *P. infestans*. As a result, a
fungus population with resistance will probably arise, and
resistance to fungicides will probably become a problem in
control of late blight.

Many protective fungicides control late blight effectively and economically. Most are applied at regular intervals of
5, 7, or 10 days depending on the temperature, moisture conditions, and the proximity of late blight in the growing area.

Various systems have been developed for accurate timing
of fungicide applications to control *P. infestans* using temperature, rainfall, and/or relative humidity (46). Such systems were used for more than 13 years in the northeastern
United States, but were not widely accepted by growers because
they were not available on a timely, regular, and localized
basis (47). A model system known as Blitecast (47) was developed in Pennsylvania using the proven systems. For Blitecast
a computer program was written to forecast late blight occurrence and to schedule fungicide applications. The system has
been tested in Pennsylvania (47) and New York (48) and shown
to forecast late blight occurrence correctly and reduce the
number of spray applications to the minimum necessary to control the disease. Potato growers in Pennsylvania are using
the Blitecast system, and in Maine a modified version of
Blitecast is available and is being used by growers there.

Fry (17,48) has studied the effect of general resistance
to *P. infestans* in different potato varieties in relation to
reducing fungicide applications. First, he investigated a
method of quantitating polygenic resistance. For late blight,
polygenic resistance reduced the rate of epidemic development
as did the periodic application of a protectant fungicide.
The approach was to prepare a standard curve of the rate of
epidemic development versus fungicide dose. The rate of epidemic development for cultivars with polygenic resistance was
less than that for cultivars with less polygenic resistance.
The effect of polygenic resistance could be equated with a
given dose of fungicide. Thus, the research reveals that the
fungicide dose can be safely reduced on cultivars with polygenic resistance. On the basis of reduction in apparent infection rate (r), the effect of polygenic resistance in Sebago
was equivalent to approximately 0.5 pounds of fungicide/ha
applied weekly to Russet Rural, the susceptible clone (17).

Secondly, Fry (48) has also shown that the use of cultivars with polygenic resistance can be combined with forecasting procedures to reduce fungicide use.

Fungicides were somewhat more effective in controlling
P. infestans if applied according to "Blitecast" (47) than if
applied after each 1.25 cm of rain (49), and the most efficient use of fungicide was to apply a reduced level of

fungicide according to the "Blitecast" system.

Insecticides

Without insect control, potatoes in many areas of the
United States would probably produce no yield at all. Average
potato yields in the United States rose dramatically with the
introduction of DDT in the 1940s. In 1915-1919 the average
hundredweight (cwt) per acre yields in the United States were
55.7 cwt (50). In 1935 they were only 66 cwt/acre but by
1945 had risen to 94 cwt/acre. Since then they have steadily
increased, i.e., 1955—162 cwt/acre, 1965—210 cwt/acre (51).
In 1976 average yields were above 250 cwt/acre. Much of these
yield increases are due to better cultural methods, irrigation, fertilizer practices, and weed and disease control, but
a significant share of the increase is due to insect control,
primarily with organic pesticides. Pimentel et al. (52) cited
figures showing that losses in yield and quality from potato
insects have declined from 22% in 1910-35 to 16% in 1942-51,
to 14% in 1951-60, to 10-12% in 1975.

Potato insects rapidly became resistant to DDT and other
organochlorine insecticides, and increasingly toxic compounds
such as organophosphates and carbamates are being utilized to
control potato insects. The entomological literature has in-
numerable examples illustrating the genetic plasticity of
insect populations (53). In addition, Peterson (54) in 1963
showed that insecticide applications with DDT and other insecticides often result in resurgence or considerably higher
populations of the green peach aphid *(Myzus persicae)*.

Prior to the use of DDT, chemicals were also used to control insects. As a boy I can still recall using Paris Green
(which contained lead arsenate) to dust potatoes for Colorado
potato beetles. However, considerable insect control was
achieved through cultural practices, mechanical measures, and
biological control prior to the 1940s.

What is needed to avoid the well-documented problems of
insect populations resistant to insecticides, undesirable
residue levels in food, unfortunate effects on wildlife, rapid
resurgence of target pest populations following treatment,
outbreaks of unleashed secondary pests, and the obvious hazard of extremely toxic chemicals to farmers? I think the
answer for the potato industry is obvious, namely, an IPM
program for potatoes. This will call for greater investments
in research on producing potato varieties resistant to insects. Cultural methods of potato insect control such as
sanitation practices, tillage, rotations, time of planting,
trap-cropping, effect of fertilizers, destruction of weeds and

other alternate hosts, crop spacing, harvesting procedures, water management, etc., need to be more intensively studied. Biological control and the use of chemicals to prevent maturation or sexual reproduction of potato insects have been given little attention. Reduction of "cosmetic standards" for potatoes as suggested by Pimentel et al. (52) could reduce insecticide use significantly.

Progress has been made in Maine in insect pest management, specifically for the control of green peach aphid to reduce losses due to potato leaf roll virus (PLRV)[3]. Some program components are: (1) aphid elimination in home gardens and on bedding plants and transplants coming to Maine from other states and local greenhouses, (2) roguing schools to train certified seed growers to eliminate PLRV plants early in the season, (3) aphid trapping in growers' plots to give advance warning of when winged green peach aphid flight was imminent, and (4) aphid traps (yellow pans) located throughout potato growing areas, to determine green peach aphid migration. This extension program disseminates information to growers concerning timely applications of foliar sprays, by newsletters, postcards, television, and radio. A pest management program in Idaho (36) has similar components.

CONCLUSIONS

IPM of potatoes implies that entomologists, plant pathologists, nematologists, and weed scientists work together with their colleagues in other disciplines to develop strong pest management systems. Such systems must be integrated with the entire package of management practices used in potato production.

The compartmentalization of disciplines concerned with the several classes of pests into departments, programs, divisions etc., of plant pathology, entomology, nematology, or weed science, means that many studies important to pest management are being neglected. We mark our territorial boundaries, and he who trespasses, beware! Unfortunately, the pests pay no attention to our boundaries. Thus studies of vector-virus relationships suffer; we duplicate our insufficient efforts in monitoring the movement of pests, but our efforts to breed crops with multiple resistance to diseases, insects, and nematodes often never get beyond the planning stage.

In a particular state or region where sufficient potato

---

[3]*F.R. Holbrook, personal communication*

acreage exists to make an IPM program feasible, potato researchers, seed certification personnel, extension workers,
and representatives from the private and public sector who are
involved in potato production need to initiate a continuing
dialogue and to establish working groups in organizing potato
pest management programs. Care must be taken to assure that
adequate representation of all pertinent potato crop protection, improvement, and production disciplines are available
to the program to assure that the recommendations of one discipline do not nullify or contradict those of another.

It will be essential for administrators, at least initially, to provide adequate funding for such activities and
to identify leaders who can devote full time to coordination
and implementation. Researchers or extensionists cannot be
expected to initiate such programs in addition to their already full-time activities. In this paper I have attempted
to show that many of the components of a strong program of IPM
already exist and that many are being utilized by potato producers. A strong research base exists for pest management
programs, but implementation will rapidly demonstrate that
additional research is needed. The components—cultural
practices, host resistance, biological control, legal regulations, and pesticide use—generally are being utilized as
separate bits and pieces, not as a coordinated, comprehensive
plan of integrated potato pest management. To put all of the
components together, it is essential that multidisciplinary
teams be formed. Teams can occasionally be formed by strong
individual leaders, but usually the teams will have to be organized by strong-willed administrative decisions.

Throughout this paper areas needing additional research
funding have been stressed. However, it is not researchers,
but rather extension personnel who will ultimately bring pest
management systems to potato producers. It should be emphasized that they should be involved at all stages in the planning and implementation of research to develop pest management
programs.

Extension personnel who are convinced of the efficacy of
the pest management program and who can relay this information
with enthusiasm to key growers are a final step in the implementation of a pest management program.

REFERENCES

1. Andrilenas, P.A. 1974. Farmers' use of pesticides in
   1971—Quantities. U.S. Dept. Agr., Econ. Res. Serv., Agr.
   Econ. Rep. 252.

2. ———. 1975. Farmers' use of pesticides in 1971. Extent of crop use. U.S. Dep. Agr., Econ. Res. Serv., Agr. Econ. Rep. 268.

3. Harlan, J.R. 1976. The plants and animals that nourish man. Sci. Am. 235(3):88-97.

4. United States Department of Agriculture. 1960. Index of Plant Disease in the United States. Agr. Handb. No. 165, Washington, D.C.

5. Akeley, R.V. 1966. Current status of potato breeding in the United States. Proc. 3rd Triennial Conf. Europ. Assoc. Potato Res. pp. 113-26.

6. Anonymous. 1975. Entered acreage of certified seed potatoes by varieties 1975. Certification Sect. Potato Assoc. Am., Am. Potato J. 52:365-72.

7. Webb, R.E. 1976. Notice to potato seed growers of the release of Atlantic, a new potato variety combining high quality, pest resistance, and wide adaptation. Am. Potato J. 53:428-30.

8. Johansen, R.H., J.T. Schulz, J.E. Huguelet. 1969. Norchip, a new early maturing chipping variety with high total solids. Am. Potato J. 46:254-58.

9. Johansen, R.H., J.T. Schulz, J.E. Huguelet. 1969. Norchief, a new smooth type, high total solids, red-skinned potato variety. Am. Potato J. 46:298-301.

10. Anonymous. 1974. International Research in Agriculture. Consult. Group in Int. Agric. Res., New York.

11. Goodrich, C.E. 1863. The potato, its diseases, with incidental remarks on its soils and culture. N.Y.S. Agr. Soc. Trans. 23:103-34.

12. Plaisted, R.L. 1971. 400 years of potato evolution. N.Y. Food Life Sci. 4(2 and 3):24-26.

13. Simmonds, N.W. 1964. Studies of the tetraploid potatoes. II. Factors in the evolution of the Tuberosum group. J. Linn. Soc. London (Bot.) 59:43-56.

14. Plaisted, R.L., H.D. Thurston, W.M. Tingey. 1975. Five cycles of selection within a population of S. tuberosum spp. andigena. Am. Potato J. 52:280. (Abstr.)

15. Thurston, H.D. 1971. Relationship of general resistance: late blight of potato. Phytopathology 61:620-26.

16. Niederhauser, J.S. 1968. Resistance to Phytophthora infestans in Mexico. First Int. Congress of Plant Pathology. Imperial College, London. p. 138. (Abstr.)

17. Fry, W.E. 1975. Integrated effects of polygenic resistance and a protective fungicide on development of potato late blight. Phytopathology 65:908-11.

18. James, W.C., C.S. Shih, L.C. Callbeck. 1973. Survey of fungicide spraying practice for potato late blight in Prince Edward Island. 1972. Can. Plant Dis. Surv. 53:161-66.

19. Gibson, R.W. 1971. Glandular hairs providing resistance to aphids in certain wild potato species. Ann. Appl. Biol. 68:113-19.

20. Granovsky, A.A. and A.G. Peterson. 1954. Evaluation of potato leaf injury caused by leafhopper, flea beetles, and early blight. J. Econ. Entomol. 47:894-902.

21. Radcliffe, E.B. and F.I Lauer. 1970. Further studies in resistance to green peach aphid and potato aphid in the wild tuber-bearing *Solanum* species. J. Econ. Entomol. 63:110-14.

22. Radcliffe, E.B. and F.I. Lauer. 1971. Resistance to green peach aphid and potato aphid in introductions of wild tuber-bearing *Solanum* species. J. Econ. Entomol. 64:1260-66.

23. Sanford, L.L. and J.P. Sleesman. 1970. Genetic variation in a population of tetraploid potatoes; response to the potato leafhopper and the potato flea beetle. Am. Potato J. 47:19-34.

24. Shalk, J.M., R.L. Plaisted, L.L. Sanford. 1975. Progress report: resistance to the Colorado potato beetle and potato leafhopper in *Solanum tuberosum* subsp. *andigena*. Am. Potato J. 52:175-77.

25. Sleesman, J.P. 1940. Resistance in wild potatoes to attack by the potato leafhopper and the potato flea beetle. Am. Potato J. 17:9-12.

26. Sleesman, J.P. and F.J. Stevenson. 1941. Breeding a potato resistant to the potato leafhopper. Am. Potato J. 18:280-98.

27. Tingey, W.M. and R.L. Plaisted. 1976. Tetraploid sources of resistance to *Myzus persicae, Macrosiphum euphorbiae* and *Empoasca fabae*. J. Econ. Entomol. 69:673-76.

28. Shepard, J.F. and L.E. Claflin. 1975. Critical analyses of the principles of seed potato certification. Annu. Rev. Phytopathol. 13:271-93.

29. Jones, E.D. and K.W. Knutson. 1976. Sources of financial support for seed potato certification programs and offical seed potato farms in the U.S. and Canada. Am. Potato J. 53:31-42.

30. O'Brien, M.J. and A.E. Rich. 1976. Potato Diseases. U.S. Dep. Agr., Handb. No. 474.

31. Brodie, B.B. 1976. Managing population densities of *Heterodera rostochiensis*. J. Nematol. 8:280. (Abstr.)

32. Walker, J.C. 1952. Diseases of Vegetable Crops. McGraw-Hill, New York.

33. Lapwood, D.H. 1966. The effects of soil moisture at the time potato tubers are forming on the incidence of common scab *(Streptomyces scabies)*. Ann. Appl. Biol. 58:447-56.

34. Shands, W.A., G.W. Simpson, H.J. Murphy. 1972. Effects of cultural methods for controlling aphids on potatoes in northeastern Maine. Univ. of Maine (Orono) Tech. Bull. 57.
35. Shands, W.A. and G.W. Simpson. 1969. Bioenvironmental control of the green peach aphid, *Myzus persicae*. Am. Potato J. 46:56-58.
36. Cunningham, G.L. and L.E. Sandvol. 1975. Idaho green peach aphid pest management program—3rd Ann. Rept: 1974. Univ. Idaho, Misc. Ser. No. 23.
37. Adler, E.F., G.C. Klingman, W.L. Wright. 1976. Herbicides in the energy equation. Weed Sci. 24:99-106.
38. Yip, C.P. 1975. Competitive ability of potato cultivars with major weed species. M.S. Thesis, Cornell University.
39. Sweet, R.D., C.P. Yip, J.B. Sieczka. 1974. Crop varieties: can they suppress weeds? N.Y. Food Life Sci. 7(3):3-5.
40. Weinhold, A.R., J.W. Oswald, T. Bowman, J. Bishop, D. Wright. 1964. Influence of green manures and crop rotation on common scab of potatoes. Am. Potato J. 41:265-73.
41. Shands, W.A., G.W. Simpson, I.M. Hall, C.C. Gordon. 1972. Further evaluation of entomogenous fungi as a biological control agent of aphid control in northeastern Maine. Univ. of Maine (Orono) Tech. Bull. 58.
42. Shands, W.A., G.W. Simpson, C.F. Musebeck, H.E. Wave. 1965. Parasites of potato-infesting aphids in northeastern Maine. Maine Agr. Exp. Sta. Tech. Bull. T 19.
43. Shands, W.A., G.W. Simpson, H.E. Wave, C.C. Gordon. 1972. Importance of arthropod predators in controlling aphids on potatoes in northeastern Maine. Univ. Maine (Orono) Tech. Bull. 54.
44. Dekker, J. 1972. Resistance. *In* Systemic Fungicides. R.W. Marsh, ed. Longman, London.
45. Georgeopoulos, S.G. and C. Zaracovitis. 1967. Tolerance of fungi to organic fungicides. Annu. Rev. Phytopathol. 5:109-30.
46. Krause, R.A. and L.B. Massie. 1975. Predictive systems: modern approaches to disease control. Annu. Rev. Phytopathol. 13:31-47.
47. Krause, R.A., L.B. Massie, R.A. Hyre. 1975. Blitecast: computerized forecast of potato late blight. Plant Dis. Rep. 59:95-98.
48. Fry, W.E. 1977. Integrated control of potato late blight—effects of polygenic resistance and techniques of timing fungicide applications. Phytopathology 67:415-20.
49. Barriga, O.R., H.D. Thurston, L.E. Heidrick. 1961. Ciclos de aspersion para el control de la "gota" de la papa. Agr. Trop. 17:617-22.
50. Stuart, W. 1923. The Potato. Lippincott, Philadelphia.

51. MacFarland, C.S., Jr. (ed.) 1969. American Potato Yearbook. 22:64–65.
52. Pimentel, D., E.C. Terhune, W. Dritschilo, D. Gallahan, N. Kinner, D. Nafus, R. Peterson, N. Zareh, J. Misiti, O. Haber-Schaim. 1977. Pesticides, insects in foods, and cosmetic standards. BioScience 27:178–85.
53. Georghiou, G.G. 1972. The evolution of resistance to pesticides. Annu. Rev. Ecol. Syst. 3:133–67.
54. Peterson, A.G. 1963. Increases of the green peach aphid following the use of some insecticides on potatoes. Am. Potato J. 40:121–29.

DISCUSSION

J. COX: Are there fertilization recommendations for control of some pests?
H.D. THURSTON: Yes. For example, plowing soybeans under, or the use of "green manure" crops, controls some pests, but we don't know how this works. There are specific recommendations; for example, to control scab, you can manipulate the soil pH. However, there is no comprehensive program of potato pest management recommendations.
E. JANSSON: Have you encountered an economic bind on research on IPM because of poor funding?
H.D. THURSTON: Yes we have, a few times. Insects on potatoes are generally controlled using pesticides, which then leads to resistance. Against the Colorado potato beetle, for example, growers on Long Island have to change chemicals about every 4-6 years. Because of this, and other problems with pesticides, there is a general trend to look for alternatives. We need more research in order to develop and implement alternatives.
M. JACOBSEN: You have mentioned that there are "bits and pieces" of IPM research available that could be applied. What interest has USDA shown in applying them, and what aggressiveness have they shown in implementing this research?
H.D. THURSTON: Much of it is being applied. But often the recommendations from one discipline contradict those of another, and there are not large-scale programs to integrate these. Would someone from USDA like to comment?
W. KLASSEN: Implementation is through Cooperative Extension, and research is conducted at agricultural experiment stations and federal research stations. These are coordinated, but there are problems getting basic research and components of control together.

M. JACOBSEN: For potatoes alone, we spend a lot on pesticide applications, but how much is spent on IPM?
J. GOOD: For all crops, we have $4.4 million slated to put IPM teams together; we hope there will be a coordinator in each state this year. We need to work on specific insect and disease programs, and then put them together. We're just starting to do that.

INSECT CONTROL IN CORN—PRACTICES AND PROSPECTS

William H. Luckmann

Illinois Natural History Survey and
Illinois Agricultural Experiment Station
Urbana, Illinois

Corn is the principal cash grain crop of the United
States. In 1976, 71 million acres were harvested, producing
6.2 billion bushels of grain at a value of $14.4 billion[1].
Acres of corn grown for grain lead by a wide margin corn grown
for silage, processing, fresh market, and popcorn. Acreages
vary, with Iowa planting approximately 13 million acres annually and Alaska reporting less than 5 acres[2]. Corn (maize) is
ranked as the third most important cereal food crop of the
world[3].

   This report, while covering corn pests in general, concerns mainly the central portion of the United States where
most corn is grown. The five Corn Belt States, Illinois,
Indiana, Iowa, Missouri, and Ohio, plant approximately 55% of
the corn acreage, and Iowa and Illinois account for one-third
of all corn grain produced in the United States. The Corn
Belt States are top users of insecticides and herbicides and
most is applied to corn (1).

PESTS AND CONTROL PRACTICES IN CORN

Status of Pests

   A survey of 35 state recommendations for pest control in

---

[1]*Department of Agricultural Economics, University of
Illinois, personal communication*
[2]*R.H. Washburn, University of Alaska, Palmer, Alaska,
personal communication*
[3]*G.F. Sprague, Department of Agronomy, University of
Illinois, personal communication*

corn, excluding stored grain insects, shows controls are
recommended for 34 insect pests or pest groups in the United
States (2). At some time, or at some locale, probably most
of these pests are capable of causing a catastrophic event
within a field or group of fields, with considerable loss of
yield or outright destruction of the crop. Other pest insects
are not so restricted in time and space and annually infest
millions of acres, reducing yield by a small or by a significant amount. A few are very destructive, but because they
are restricted in distribution and are not found in major
corn-producing areas have less impact on the national crop.
Pest species such as the western corn rootworm, northern corn
rootworm, European corn borer, fall armyworm, black cutworm,
and corn leaf aphid, appear to monopolize the corn scene,
causing greatest concern to growers and greatest use of insecticides on corn. Some of these species, because of high
pest density, cause serious damage on an annual basis; others
have the potential to cause serious damage, though damage may
not occur annually. Most of the major pests are widely distributed or disperse into areas where most corn is grown.

The list of insect pests of corn compiled in the survey
of state recommendations is presented in Table 1. The pests
are listed by common name using the "Common Names of Insects"
approved by the Entomological Society of America (3). Use of
approved common names was not always possible. The designation "wireworms," for example, includes many species, with
certain ones more important in some areas than in other areas.
Further, I have categorized the insects as major and consistent pests, major but sporadic pests, and pests of moderate to
minor importance, based on annual destructiveness, potential
to be destructive, and on distribution or dispersal into areas
where much corn is available for attack and economic loss. I
have solicited the advice of several colleagues in categori-
zing these pests, to make the listing as accurate as possible
on a national basis.

Control Practices

Corn is protected from insect pests mainly by use of
insecticides, by cultural practices, and by insect-resistant
varieties, in that order. Biological control in the form of
man-manipulated parasites, predators, and pathogens is not
currently practiced. Other controls, such as light traps,
pheromones, and release of sterile males are not now used in
corn.

Resistant varieties contribute significantly to control
of the European corn borer, but plants resistant to the corn
borer during the vegetative whorl stage of development, when

Table 1. Compilation of the Insect Pests of Corn Obtained
from Control Recommendations Issued by State Cooperative Extension Services in the United States

Major and consistent pests[a]	Pests of moderate to minor importance[a]
Western corn rootworm	True armyworm
Northern corn rootworm	White fringe beetle
European corn borer	Grasshoppers
Corn earworm	Corn flea beetle
Fall armyworm	Chinch bug
	Southern corn rootworm
	Lesser cornstalk borer
Major but sporadic pests[a]	Western bean cutworm
	Other cutworms[b]
Black cutworm	Seed corn beetles
Corn leaf aphid	Seed corn maggot
Southwestern corn borer	Banks grass mite
Wireworms	Two-spotted spider mite
	Billbugs
	White grubs
	Common stalk borer
	Garden symphylan
	Japanese beetle
	Sod webworm
	Grape colaspis
	Thrips
	Dusky sap beetle
	Stink bugs
	Southern cornstalk borer
	Corn root aphid

[a]Pests categorized by W.H. Luckmann.
[b]Sandhill cutworm, variegated cutworm, glass cutworm,
dark-sided cutworm, dingy cutworm, bronzed cutworm, bristly
cutworm, spotted cutworm (4).

the first corn borer generation occurs, are usually not resistant later during the flowering and fruiting stages, when the
second and third generations of the corn borer attack the
plants. Similarly, plant material tolerant to the southwestern corn borer and the corn leaf aphid, and perhaps others,
aids in control, but resistant plants alone are not available
for control of any insect pests of corn. However, hybrids
do vary in degree of susceptibility, and it is not likely that
any hybrid corn very susceptible to damage from the European
corn borer or other major corn pest would be released by

commercial seed companies.

Both chemical and mechanical plant factors act as a basis for resistance to corn pests. The chemical factor, DIMBOA, is largely responsible for resistance of whorl stage corn to feeding by European corn borer larvae. Tightness of ear husk, length of husk, and quantity of silks in the silk channel mechanically restrict the amount of damage caused by the corn earworm and possibly the fall armyworm, European corn borer, and nitidulid beetles. In addition, tightly overlapping husks offer no protection for corn leaf aphids, which readily colonize and feed under the husks of loose, open ears.

Cultural insect controls consist of crop rotation, timing of planting, plowing to eliminate weeds or to make plant debris unsuitable for overwintering larvae, and control of weeds that attract and harbor insect pests. However, any assessment of contemporary insect control practices in corn will reveal that controls are almost exclusively chemical, and it must be assumed that recommendations reflect practices. For the most part, control with insecticides is effective and reliable.

Soil insecticides are widely used as prophylactic treatments for soil insects. Most are applied as granules in a 7 in. band over the row at time of planting. Some soil insecticides that are not phytotoxic are applied as granules in the furrow at planting time. Only a small percentage of acres are treated with broadcast applications of insecticides, which require several times the amount normally used in band or furrow applications. Seed treatments are widely used, with the insecticide normally applied as a planter box treatment at time of planting. Sprays, dusts, and granules are recommended for foliage and stalk-feeding insects, usually to be applied on a use-as-needed basis.

DDT, aldrin, dieldrin, and heptachlor were the principal corn insecticides a decade ago, but the less persistant organophosphate and carbamate insecticides currently predominate the control scene. The cyclodiene insecticides are either banned or their use is restricted, but in any case many of the soil insects for which these compounds were used have developed resistance. The less persistent organophosphate and carbamate insecticides have not provided acceptable control for a few corn pests. Cutworms are frequently cited as an example, and though the black cutworm outbreaks of 1977 and recent years have been falsely attributed to the unavailability of aldrin and heptachlor, it is reasonable to conclude that there would have been less damage if these insecticides were available.

Some of the insecticides that previously gave good

control are performing poorly or providing less than satisfactory control in some fields. This is especially true of
the carbamate soil insecticides, metalkamate (Bux-ten) and
carbofuran (Furadan), where either of these insecticides has
been used annually in the same field for two or more years.
Many states suggest rotating from a carbamate to an organophosphate insecticide or vice versa when control is unsatis-
factory with either of these classes of insecticides (2).

A guide to contemporary control and control aids for insect pests of corn is presented in Table 2. The pattern of
Xs in this tabulation readily identifies the major thrust in
control practices. Crop rotation, plowing to destroy or expose corn stubble to make it unsuitable for overwintering
pests, early planting, and use of corn hybrids tolerant or
resistant to the pest are regarded as control aids. Less than
a dozen pests or pest complexes have damage thresholds to aid
in decision making and most of these thresholds are not substantiated by research.

The recommendations and practices released by most states
contain a tabulation of insecticide recommendations plus moderate to substantial amounts of information about the major
corn pests in each state and specific directions to aid in
avoiding or controlling these pests. Special emphasis is
often given to insect problems that are likely to occur in
minimum and no-tillage systems of corn production, where pest
problems tend to be more numerous and control more difficult.
Advice is given on the reasons for early planting and how
this practice will lessen pest severity. Regrettably, these
control aids sometimes appear as footnotes and explanations
or reasons for them are lacking.

A very significant practice is the emergence of extension pest management programs and commercial pest scouting in
corn. The accomplishments of the pest management programs in
corn funded by the Federal Extension Service, USDA, have
caught the attention of the corn farmer. There is an alertness on the part of the farmer that did not exist five years
ago and while other factors (control failure, pest resistance,
outbreaks) may contribute to the interest in pest management,
the visible contribution of extension pest management program
specialists must be credited for increasing interest and concern on the part of the grower. Similarly, research in pest
management has intensified in corn in the past 5-7 years,
thanks to USEPA, USDA-ARS, state experiment station directors,
and to the CSRS Competitive Grants program. During the past
five years, more has been learned about the biology, behavior,
density, and pest distribution of the major pests of corn than
was learned in the preceding two decades. These research data
and the accomplishments of the extension pest management

Table 2. Types of Controls Used for Insect Pests Attacking Corn in the United States

Pest/Pest complex	Seed treatment	Soil insecticide	Foliage insecticide	Insecticide baits	Crop rotation	Early planting	Stalk destruction	Resistant hybrids	Specific damage threshold guidelines
Corn rootworm larvae		X			X				X[a]
Corn rootworm adults			X						X
European corn borer			X			X	X	X	X
Corn earworm			X			X			
Fall armyworm			X			X			
Cutworms		X	X	X					X
Western bean cutworm			X						X
Corn leaf aphid			X			X		X	X
Southwestern corn borer			X			X	X		X
Wireworms	X	X							
Seed beetles	X	X							
Seed maggot	X	X							
True armyworm		X	X						
White fringe beetle		X							
Grasshoppers			X						X
Corn flea beetles			X						
Chinch bug			X						
Lesser cornstalk borer		X							
Spider mites			X						X
Billbugs		X							
White grubs		X[b]							
Common stalk borer			X						
Garden symphylan		X							
Japanese beetle adults			X						
Sod webworm			X						
Grape colaspis		X[b]							
Thrips			X						
Dusky sap beetle			X						
Stink bugs			X						
Southern cornstalk borer			X				X		
Corn root aphid		X							

[a]Based on adult density of 1 or more/plant the preceding year.

[b]Heptachlor if available. Existing stocks can be used in 1977, otherwise no effective control.

programs are beginning to filter into control practices.

In summary, current practices of insect control in corn are very dependent on chemical insecticides with other controls usually in a subordinate position. There is reason to believe that insecticides will continue to be the first choice in control for some time, as research has not produced the data that will permit sole reliance on other methods. The short-term approach to corn pest control must depend on insecticides. Recent events, such as the dramatic increase in the density of insects such as rootworms and cutworms, control failure with some soil insecticides that performed well in the past, changes in agronomic practices such as very early planting, the concern for pest resistance, and the suspension by USEPA of the use of aldrin and heptachlor soil insecticides have contributed an element of uncertainty to corn pest control. S.A. Forbes (5) in his presidential address to the 21st Annual Meeting of the American Association of Economic Entomologists described in an interesting manner the variables in insect control. Much of what he said then applies today, and a portion of that address is reproduced here:

> Economic entomology is an extremely complex subject, not only by reason of the number of factors which it must include, but especially because of the variability of many of these factors, and our inability to predict the course of events with certainty in our field. We study the present and the past in a practical way in order that we may predict the future. We observe, generalize, experiment and verify in order that we may be able to say to the farmer or the fruit grower, "Do thus and so in any given case, and this or that desired result will follow;" but we can rarely express our conclusions safely in so definite a form. Often the best we can fairly say is that if the weather should be wet, or dry, or very wet, or very dry, for the last two or three or four years; or if the winter has been, or is to be, open or severe; if the crop in question has been preceded by some other kind of crop, or by one of the same kind; and if the insect situation was thus and so last year and the year before; if, furthermore, the land is light or heavy, high or low, well drained or wet; if it has had this or the other management or treatment during the last year or two; and if several other variable elements of the problem vary to such or such a degree, in this or the other direction—then if the operation X be performed, the result will *probably* be Y, but with what *degree* of probability it is impossible for us to say.

PROSPECTS AND STRATEGIES FOR PEST MANAGEMENT IN CORN

The primary tool currently available to control insect pests of corn is chemical insecticides, but the primary tool for control need not be the principal means of control. If the current interest continues and if the dollars are available to maintain the momentum begun in the past 5-7 years, pest control in corn will take on a distinctly different mode. There are several reasons unique to corn and to some of the corn pests that could allow this to happen, providing there is a will to do it and the dollars to support the effort. Several approaches, needs, and obstacles to pest management in corn are discussed in the following paragraphs.

The Zero Concept

Most corn is produced north of latitude 37°N in the United States, and the majority of fields do not have an annual insect pest problem worthy of chemical control. Thus, we must learn to measure zero or near zero, and though we may not know all that needs to be known about a pest, there is no need for use of a chemical insecticide if the pest is not there or is present in low numbers. The adoption and use of the zero concept can have a great impact on pest control in corn. The argument that we need to know everything about a pest before we can have a pest management program is not valid.

One of the objectives of the current USEPA project "Bionomics and Management of Soil Arthropod Pests of Corn" is a survey of the distribution and density of wireworms in corn in Nebraska, Iowa, Missouri, Illinois, Indiana, and Ohio (6). A bait mixture of equal parts wheat grain and corn grain, using one cup of the mixture per site at 10 sites per field in early spring, appears to be very effective in identifying fields with wireworms. The damage threshold for wireworms has not been determined, but in this survey wireworm larvae were not found in most of the fields. The data in Table 3 represent one year's observations, but they are representative of results obtained with baits in the study covering several years. A 3-year survey for wireworm larvae in Indiana from 1972 through 1974 showed 3.8% of 234 study fields contained wireworms (7).

The wireworm baiting technique can be used by farmers and insect scouts to identify fields with wireworm larvae before corn is planted. However, its greatest utility will be the identification of fields with zero or near zero numbers of wireworms. Researchers at the University of Missouri

Table 3. Summary of Preliminary Baiting[a] Studies in Six
States to Detect Presence of Wireworm Larvae Prior to Planting of Corn, 1975

	No. Sampled Counties	Fields	No. of fields with wireworms in baits	No. of baited fields with wireworm damage	% of baited fields With wireworm larvae	% of baited fields With wireworm damage
Illinois	16	124	2	1	1.6	0.8
Iowa	4	12	4	0	33.3	0.0
Indiana	5	72	10	0	13.9	0.0
Missouri	6	63	4	1	6.3	1.6
Ohio	10	52	4	0	7.7	0.0
Nebraska	3	19	7	2	36.8	10.5
	44	342	31	4	9.0	1.2

[a]The wireworm bait was an equal mixture of corn and
wheat grain using one cup per baiting site with 10 bait locations per field.

are refining the baiting technique to improve detection and
to establish a damage threshold for wireworm larvae in corn.

Pest Monitoring

Detection, measurement, and prediction techniques are
important areas of research needed in the future for pest
management (8). Such methods have been developed for only a
few corn pests. The outstanding example is the European corn
borer. Reasonably accurate methods of detection and measurement are currently being refined for such insects as the
northern corn rootworm, western corn rootworm, and wireworms.
Finite techniques may not always be necessary, as we can tolerate some damage without loss in yield. Accurate techniques
of detection and measurement could allow controlled use of
some of the insecticides that might otherwise be lost because
of environmental risk.

There is a serious lack of data on threshold levels to
guide treatment with insecticides. Research has not validated thresholds for most pests of corn. Insecticide recommendations suggest treat-as-needed, but need is not defined.
Years of experience and a "feel" for the problem guide many
research and extension people in decision making.

Everything an insect does and all the things that happen to it are in response to physical and chemical factors. Dispersal, establishment, mating, increase, survival, behavior, and control obtained with an insecticide are all regulated or influenced by extrinsic and intrinsic factors. Most or all of these factors can be measured and expressed as numbers, and numbers can be put into a computer. The computer is essential if we expect to achieve sophistication and reliability in pest management. It can be especially useful in predicting pest population growth and probability of occurrence, in meshing various kinds of controls, and in analyzing effectiveness and costs.

Pest Resistance and Insecticides

Resistance is one of the principal reasons for adopting pest management, and may well be the key to acceptance of pest management programs. The rapid buildup of high-level resistance, experienced with some corn pests with the chlorinated hydrocarbons, is not likely to occur with the switch to the less-persistent insecticides. Resistance in corn pests should develop at a slower rate and the level of resistance attained should not be as high as with the chlorinated hydrocarbon insecticides. What will happen is that control will become less effective. Reduction in the environmental pressure from insecticides will increase the utility and longevity of the products we are currently using.

The recent failure of Bux-ten and Furadan to control western and northern corn rootworm larvae, in fields where these insecticides have been used annually for several years, is disturbing. Research reports on Furadan[4] (9) suggest a change in the microflora composition in the soil in these fields leading to more rapid biological degradation of carbamate insecticides. It has been known for a long time that soil microorganisms have considerable effect on the stability of insecticides in the soil (10), but if the field observations and research reports are correct, we should have real concern regarding this phenomenon and its implications.

Pest Management/Crop Production

Energy conservation and the Federal Water Pollution

---

[4]*W.N. Bruce, Illinois Natural History Survey, Urbana, personal communication*

Control Act could have a significant impact on corn production and pest problems. State programs currently being developed under PL 92-500, Section 208, Water Quality Management Program, will almost certainly emphasize control of erosion and sedimentation through regulations requiring soil conservation and agronomic practices that prevent excess soil loss. Important changes could include increased emphasis on minimum tillage to leave more surface debris. The plans being developed will almost certainly impose changes in management of some land used for corn production and thus will affect pest problems.

In the past, crop protection recommendations have been developed independently for insects, weeds, and diseases, and crop protection policies have rarely been considered when changes are made or new crop-production systems are developed. Crop production almost invariably takes precedence over pest control. Pest control is routinely included as a crop expense, but adequate economic analysis is rare. Low cost and the reliability of pesticides has in large part relegated them to routine tools of production, whether they are needed or not, though omnipresent weeds must be reasonably well controlled or the crop is lost. Decisions on use are generally short-term and farm-oriented. This situation almost entirely explains the fantastic growth in the use of pesticides during the past quarter of a century (11). Corn producers have done very well in integrating the bits and pieces of information for short-term application, but it is essential that long-term pest control methodologies and policies are developed.

In each of the major corn-producing areas of the United States, some other annual row crop of almost equal value is grown. In the Midwest, that other crop is usually soybeans, but could also be grain sorghums, forage sorghums, small grains, and legumes such as alfalfa. Crop rotation breaks the rootworm cycle because the northern and western corn rootworms oviposit eggs in a field the year preceding larval attack and the larvae cannot survive on other crops. Rotation may not be economically advisable for all corn producers or on all land producing corn, but a high percentage of corn-soybean farmers rotate their crops on an annual basis as a routine management practice.

I have proposed the following State-Wide Pest Management Plan for corn rootworms in Illinois. It has been reviewed by approximately 20 agronomists, economists, plant pathologists, weed scientists, agricultural engineers, and entomologists, and the most repeated comment is "This is a good idea, but will a voluntary program work?" I propose that it will when the pyramiding beneficial effects of the program are

explained. The fact that many growers already rotate is
reason for optimism. Aside from rootworm control, additional
benefits of rotating include a reduced nitrogen requirement
for corn, long-term increase in corn yields, and an aid in
control of some weeds and diseases.

A STATEWIDE PEST MANAGEMENT PLAN FOR CORN ROOTWORMS IN
ILLINOIS

The Rationale

The northern corn rootworm and the western corn rootworm
are the major insect pests of corn in Illinois. The northern corn rootworm was first discovered in Colorado in 1824,
and has probably been in Illinois for that long. The western
corn rootworm entered Illinois in 1964, migrating from Nebraska and Iowa. Many fields in Illinois contain both species
of rootworms. Both have developed resistance to the chlorinated hydrocarbon soil insecticides (aldrin, dieldrin,
heptachlor, and chlordane), but several organophosphate and
carbamate soil insecticides still provide generally good
plant protection. However, the level of control obtained
with these insecticides would be greater if we had fewer
rootworms in our fields; past experience suggests that rootworms will develop tolerance or resistance to the organophosphate and carbamate soil insecticides. Thus, any practice
that will reduce the number of rootworms without increasing
the annual exposure of rootworms to insecticides will extend
and enhance the control of these pests with contemporary
insecticides. The plan described here is a voluntary but
deliberate program by Illinois farmers to reduce the overall
population of corn rootworms in Illinois.

The Facts

The northern and the western corn rootworms have only
one generation each year. The females oviposit eggs in the
soil in cornfields in August and September, and these eggs
overwinter and hatch the following June. Corn rootworm larvae live best on corn roots. Survival is very low and of no
consequence on grassy weeds. The larvae cannot survive on
roots of broadleaf weeds, broadleaf crops, small grains,
alfalfa, or sorghum. Thus, larvae cannot survive and emerge
as egg-laying adults when a crop other than corn is planted
in a field whose soil contains eggs of the northern and western corn rootworms.

The nitrogen requirement for corn would be reduced in a corn-soybean rotation. Experiments conducted at the Carthage Experimental Field comparing nitrogen requirements of continuous corn and corn following soybean indicates a soil nitrogen contribution of 30-40 lb per acre at lower rates of applied nitrogen and 20-30 lb per acre at higher rates of applied nitrogen (fig. 1). At the Elmwood Agronomy Research Center, the yield differential between continuous corn and corn-soybean continues to widen at higher rates of nitrogen application. Yields from the Morrow plots at Urbana and the Elmwood Agronomy Research Center show that corn following soybean has consistently yielded better than corn following corn (12,13).

Greater amounts of soil nitrogen following soybean are not the total explanation for higher corn yields on the Morrow and Elmwood plots since any difference in corn yields between the two cropping systems that was due to nitrogen would be expected to disappear at high rates of added nitrogen fertilizer. This did not occur, so differences in yields are apparently not due only to nitrogen available to the corn (12,13).

The physical properties of soil may contribute somewhat to the differences. The soil is more loose and friable following soybean than following corn. This occurs to such an extent that only discing is suggested when preparing a seedbed for corn following soybean. The more friable soil

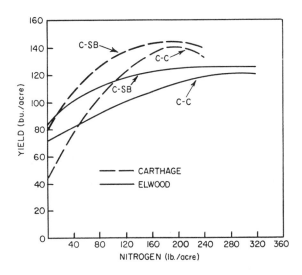

Fig. 1. Effect of crop rotation and applied nitrogen on corn yield.

following soybean could result in a better stand of corn than
when corn follows corn. This possible benefit would be important in farmers' fields only when a good seedbed was not
prepared with corn following corn (13). Rotation also allows
time for residue decomposition, which helps reduce stalk rot,
root rot, *Helminthosporium* leaf diseases, *Physoderma* brown
rot, and *Anthracnose* leaf blight of corn.

A summary of 10 years of data on crop/herbicide rotation
shows a distinct long-term yield advantage of rotating corn
with other crops regardless of the weed control system used.
An economic analysis of the 10-year experiment in Illinois
indicates that the profitability of herbicide use is very
high, ranging from $3.30 to $4.89 per $1 of herbicide cost.
This return is not affected greatly by crop sequence, and is
about the same with continuous corn as with corn-corn-soybean
and corn-soybean-wheat (14). Rotation of corn and soybean
also helps in controlling some weeds that are more difficult
to control in a continuous cropping system[5].

The Impact

Corn rootworm adults oviposit between 200 and 1,000 eggs
per female. In the fall of each year, corn rootworm eggs in
the soil of Illinois cornfields range from none to 50,000,000
eggs or more per acre. The soil in a large number of cornfields contains between 10,000,000 and 20,000,000 eggs per
acre. If only 1% of these rootworms survive to adulthood, a
field with 10,000,000 eggs per acre produces 100,000 adults
per acre, of which approximately 50,000 are ovipositing females. Using these conservative figures, a 40-acre field of
corn could produce 2,000,000 females that could lay eggs in
that field and in surrounding cornfields. Thus, by crop rotation, we can reduce by an almost incalculable figure the
number of ovipositing females on each farm in the state. Soil
insecticides perform better against a lower population of
rootworms, and many fields would require no treatment at all.
While insect-control benefits will accrue on each farm, the
impact will be greatest and most long-term when a majority of
Illinois farmers collectively recognize that they can have a
pyramiding effect in suppressing rootworms as more fields are
rotated.

[5]*E.L. Knake, Department of Agronomy, University of
Illinois, personal communication*

The Method

The program is only applicable to farmers who grow corn
and some other annual crop. In Illinois, this would mainly
apply to farmers who grow corn and soybean. The grower would
visit each of his cornfields in August, and determine which
fields had the highest number of rootworm beetles. The fields
with the highest rootworm beetle population would be deliberately planted to soybean or some other nonhost crop the following year. For example, suppose a farmer annually grows
four fields of corn and two fields of soybeans. Through field
examination, the two cornfields with the highest number of
rootworm beetles would be planted to soybeans the next year.
The other two cornfields could continue corn on corn. If the
adult rootworm count exceeded one adult/plant, a soil insecticide would be used.

Many states have descriptive directions for counting corn
rootworm beetles and for determining need for the use of a
soil insecticide. This information is contained in the
Illinois Extension Service Circular 899, "1977 Insect Pest
Management Guide, Field and Forage Crops." This circular or
specific directions for making corn rootworm beetle counts
can be obtained from county extension advisers. In Illinois,
we recommend that a soil insecticide be applied at planting in
all fields having one or more rootworm beetles per plant in
August of the preceding year. Some farmers will observe
fields containing 3-10 adults or more per plant, a high infestation. Some farmers have fields where it is difficult to
observe an adult corn rootworm and counts would not exceed one
rootworm beetle per plant. These would be considered low-
population fields. Following the example above, a farmer with
four fields of corn may find that one field has less than one
beetle per plant and the other three may have more than one
beetle per plant with two of the fields with a population of
four beetles or more per plant. The object is to survey the
fields, select the ones with the highest beetle count, and
deliberately plant those fields to soybean the following year.

The Commitment

The official Illinois recommendations given in Circular
899 state that crop rotation is the most effective method of
preventing corn rootworm damage and if feasible, corn should
not be grown two years in succession in the same field. The
plan presented here supports that recommendation, but goes a
step further. It urges a statewide awareness and acceptance
of a program that will benefit every corn farmer. It is a

long-term voluntary commitment to reduce the overall corn
rootworm population in the state, to reduce damage, and to
make control easier with insecticides.

This plan cannot be implemented by farmers who only grow
corn, but they also will benefit from the program. The plan
should not be followed if it is detrimental to erosion control
practices on a farm, but many farmers already rotate corn and
soybean as a routine production practice. On an annual basis,
prices may favor planting more corn than soybean, but it is
likely that a large number of Illinois farmers will produce
both corn and soybean and that a voluntary but planned annual
rotation program is feasible and will be profitable. Collectively, Illinois farmers have an opportunity to initiate a
program that will be of tremendous benefit to them and to all
Illinois agriculture.

OBSTACLES TO ACCEPTANCE AND USE OF PEST MANAGEMENT

There are several obstacles to pest management, but two
dealing with people are probably the most important. The
first is acceptance and use by the grower. The second is acceptance and use by the people doing pest control research and
making pest control recommendations. I believe very strongly
that management is people-oriented. Successful implementation
of pest management programs will depend largely on influencing
the people who control the pest.

Interestingly, a lack of pest problems is a hindrance to
acceptance of insect pest management by the corn farmer.
Farmers participating in the extension pest management programs in Illinois, whether on a free or fee basis, expect to
be told that they have a problem. They express reluctance to
pay for a service that tells them to take no action most of
the time. Unlike some crops where multiple applications are
needed, most pest problems in corn can be solved with a single application of an insecticide. Thus, in corn the pest
situation is usually treat or no-treat, with a high percentage
of corn acres treated only once and often with a prophylactic
application of soil insecticide as insurance against damage.
Speciality crops such as market and canning sweet corn usually have a number of pest problems annually requiring multiple
treatments and producers of these crops are very alert to pest
management recommendations.

Corn farmers are also reluctant to do their own monitoring for pests, preferring to depend on annual prophylactic
treatments or on timely news releases about the pest situation
and advice from extension personnel. Simply stated, corn
farmers have not seemed interested in going into the fields

and doing pest management monitoring themselves. However, in the past year or so, I have detected an alertness that did not exist before, and though this may be stimulated in part by recent pest outbreaks and rumors of control failure, the extension pest management programs must be credited with a large part of the interest we now see demonstrated by farmers. Note the response in acres treated shown in Table 4. In 1973, before predictions were made on number of acres needing treatment, 78% of the corn acres in these study areas were treated with a soil insecticide for rootworm larvae. These action programs are visible, and some of the knowledge obtained over the past 5 years is beginning to appear in pest control recommendations. Credit must also be given to the multi-state research projects begun 5-7 years ago. A continued momentum of these kinds of extension and research programs is essential. The decisive period for grower acceptance of pest management is the next 5-10 years.

Table 4. Acres of Corn Treated with a Soil Insecticide for Northern and Western Corn Rootworm Larvae in Intensively Monitored Pest Management Study Areas[a] in Illinois in 1974 and 1975

1974			1975		
Acres of corn	% Acres needing treatment	% Acres treated	Acres of corn	% Acres needing treatment	% Acres treated
9,004	19	67	21,756	11	57

[a]Study areas located in Boone, Hancock, Warren, and Shelby counties.

"Uncertainty," much as S.A. Forbes described it in his address 69 years ago, is overwhelmingly accepted by pest control research and extension personnel as an obstacle to pest management. The argument is if we make a single mistake in pest management, growers will no longer accept it. However, dozens of corn farmers annually experience control failure with insecticides and herbicides also. We rely on chemical control because a large amount of uncertainty is removed when we recommend insecticides. They can be used as prophylactic treatments before the fact or as emergency treatments after the fact with the belief that by using them, we have done the best science can provide. Other controls are suspect of being less effective and not reliable.

Much uncertainty about other controls, economic injury levels, economic thresholds, and pest detection/measurement/

prediction can be removed by research that truly addresses the needs of pest management. Some research very applicable to insect pest management in corn is being done and the results are near publication. The prospects for future accomplishment are excellent, provided the dollars are forthcoming to support the effort. The crucial time frame is the next 10 years.

REFERENCES

1. von Rümker, R. 1974. Farmer's Pesticide Use Decisions and Attitudes on Alternate Crop Protection Methods. Office of Pesticide Programs, Office of Water and Hazardous Materials, Environmental Protection Agency, Washington, D.C.
2. Insect Control Recommendations for Corn. 1977. Cooperative Extension Service. AR, ND, SD, NY, MS, GA, MN, NJ, AK, SC, IN, MO, FL, CO, AZ, OK, IA, VA, LA, MD, MI, KS, OH, PA, NB, NC, ID, RI, TX, ME, VT, MA, DE, KY, IL.
3. Anderson, D.M. 1975. Common Names of Insects (1975). Spec. Publ. 75-1, Entomol. Soc. Am.
4. Rings, R.W. and G.J. Musick. 1976. A pictorial field key to the armyworms and cutworms attacking corn. Res. Circ. 221, Ohio Agr. Res. Dev. Center, Wooster, Ohio.
5. Forbes, S.A. 1909. Aspects of progress in economic entomology. J. Econ. Entomol. 2(1):25-35.
6. Environmental Protection Agency. 1975. Bionomics and management of soil arthropod pests. Second Annual Report. Grant No. R802547.
7. Turpin, F.T. 1977. Wireworms and their control. 29th Illinois Custom Spray Operators Training School. Univ. Ill., Coop. Ext. Serv.
8. National Academy of Sciences. 1975. Pest Control: An Assessment of Present and Alternative Technologies. Vol. II. Corn/Soybeans Pest Control. Washington, D.C.
9. Williams, I.N., H.S. Pepin, M.J. Brown. 1976. Degradation of carbofuran by soil microorganisms. Bull. Environ. Contam. Toxicol. 15:244-50.
10. Lichtenstein, E.P. 1970. Fate and movement of insecticides in and from soils. *in* Pesticides in the soil: ecology, degradation and movement. International Symposium on Pesticides in the Soil. Michigan State University, East Lansing, Michigan.
11. Petty, H.B. and R.T. Huber. 1972. Corn insect pest management. pp. 107-119 *in* Implementing Practical Pest Management Strategies. Proc. Natl. Ext. Insect-Pest Management Workshop, Coop. Ext. Serv., Purdue University. West Lafayette, Indiana.
12. Graffis, D.W., R.G. Hoeft, E.L. Knake, D.G. McClure, M.D.

McGlamery, T.R. Peck, W.O. Scott, M.D. Thorne, J.C. Siemans. 1976. Illinois Agronomy Handbook 1977-78. Univ. Ill. Coop. Ext. Serv. Circ. 1129.
13. Welch, L.F. 1977. Soybeans good for corn. Soybean News 28(3):1.
14. Hawkins, D.E., F.W. Slife, E.R. Swanson. 1977. Economic analysis of herbicide use in various crop sequences. Ill. Agr. Econ. 17(1):8-13.

DISCUSSION

M. SAVOS: Are you talking strictly about corn for grain, or are you including sweet corn?
W.H. LUCKMANN: Sweet corn is more of a speciality crop, and my statement that most corn only needs to be treated once would not apply to sweet corn. Much of what I've said would apply, but mostly our concern is field corn grown in the grain belt.
M. SAVOS: The bulk of corn grown in Connecticut is for feed for the farmer's animals. They can't rotate because they need the corn.
W.H. LUCKMANN: We're saying rotate where you can, and we realize that this will not be everywhere.
J. GOOD: You can sell the farmer on rotations by talking about the multiple benefits: for example, organic matter is added to the soil; also some soybean problems can be controlled by rotations.
W. LOCKERETZ: How general is our knowledge of the effect of insecticides on soil microbes?
W.H. LUCKMANN: There has been a lot of work done on toxic effects of insecticides on soil microorganisms. It has long been known that some soil microorganisms break down insecticides. We find sometimes that an insecticide that *was* effective, suddenly no longer is because the microbes are breaking the insecticides down faster.

PROGRESS IN INTEGRATED PEST MANAGEMENT OF SOYBEAN PESTS

L. Dale Newsom

Center for Agricultural Sciences and Rural Development
Louisiana State University
Baton Rouge, Louisiana

INTRODUCTION

Increasing demand for protein and oil by a rapidly expanding human population has led to a substantial increase in soybean production in the United States. As recently as 1961 about 75% of the total acreage planted to soybean was located in 12 Midwest and North Central states. Since then, soybean acreage in the United States has more than doubled. Expansion of soybean production in southern states has been especially dramatic, as much as ten-fold. In Louisiana, for example, the acreage devoted to soybean production for grain increased from less than 200,000 acres in 1961 to 2,400,000 in 1977. Based on current estimates (1), 55,698,000 acres will be planted to soybean in the United States during 1977. It appears that greater increases are likely in the future (Table 1).

Fortunately, in areas of traditional production, insect pest problems are of little importance. There are no serious insect pest problems in these areas, where almost half of the total U.S. acreage is presently located and few of the major insect pests of soybeans occur there. Furthermore, the crop has been produced on a large scale for so long in the area, it appears doubtful that any native species is likely to adapt to it.

The pest situation is quite different in the southern United States. There, a large complex of important and potentially important species attacks the crop (Table 2). Additional species appear to be adapting to soybean, although significant pests have not yet developed. The challenge to the entomological profession is to prevent the development of new key pests, especially in areas where soybean is at comparatively high hazard from pests.

Table 1. Trends in Acreage Planted to Soybean for Grain in
Midwestern and Southern United States During the Period
1950-1977 (1,2)

1950-59	1961	1971	1977
Twelve midwestern states (1000 acres)			
14,434	20,203	27,594	33,765
Eleven southern states			
3,245	6,370	14,375	20,325
Total United States			
18,045	27,340	42,790	55,678

It is frightening to contemplate the problems that would
be created should the profitable production of soybean ever
require as much use of conventional chemical insecticides as
that commonly required, for example, for control of insect
pests of cotton. First, the economics of soybean production
would preclude such heavy use, and secondly, environmental
pollution would be enormously increased. Much of the soybean
acreage in the South, which is at comparatively high hazard
from insect pests, is also more sensitive to the adverse
effects of heavy insecticide use than is the case with areas
devoted to cotton production. A high percentage of the acreage now planted to soybean in most of the southern states
consists of land recently cleared of forests, permanent pastures, and temporary grazing lands. Such areas are usually
closer to streams, lakes, ponds, and woodlands than is cotton
acreage. Thus, they are more sensitive to the adverse effects
of heavy insecticide usage. Also, soybean ecosystems are
heavily used by a large variety of animals, including many
species of upland game birds, mammals, and migratory water-
fowl. Many species of prime game birds and mammals—deer,
rabbit, duck, quail, and mourning dove—use soybean extensively for food or cover. Soybean also provides excellent
habitat for large complexes of predatory and parasitic
insects.

The severe problems that have arisen from excessive use
of conventional insecticides for control of cotton pests
clearly show that development of integrated pest management
(IPM) systems requiring minimum use of conventional

insecticides provides the only reasonable approach to the control of soybean insect pests.

ASSESSMENT OF INSECT PEST PROBLEMS OF SOYBEAN

Based on planting intentions announced by growers for 1977, more than 55 million acres will be planted to soybean in the United States. For a crop of this size, it may seem surprising that insect problems have attracted relatively little attention. In a recent review article on "Soybean

Table 2. Species of Important Insect Pests That Attack Soybean in the Southern United States Listed According to the Part of the Plant Attacked

---

Roots and Nodules

Ceratoma trifurcata (Forster)
Diabrotica balteata (Le Conte)
Colaspis brunnea (Fabricius)
Rivellia quadrifasciata (Macquart)

Stems

Spissistilus festinus (Say)
Dectes texanus (Le Conte)
Elasmopalpus lignosellus (Zeller)

Foliage

Anticarsia gemmatalis (Hubner)
Pseudoplusia includens (Walker)
Plathypena scabra (Fabricius)
Epilachna varivestis (Mulsant)
Ceratoma trifurcata (Forster)
Diabrotica balteata (Le Conte)
Heliothis zea (Boddie)

Fruit

Nezara viridula (Linnaeus)
Acrosternum hilare (Say)
Euschistus servus (Say)
Heliothis zea (Boddie)
Ceratoma trifurcata (Forster)

Entomology" that treated the literature through 1975,
Turnipseed and Kogan (3) cited only 264 articles. Of these,
only 188 dealt with soybean insect pests in the United States,
their natural control agents, or the diseases that they
transmit. More than half of those cited were published during
the period 1970 to 1975. A major reason for such a comparatively small research effort devoted to soybean insect pests
in the United States prior to the last decade is that a high
percentage of the crop was being produced in the Midwest and
North Central States where there is little hazard from insect
pests.

The large increase of acreage planted to soybean in areas
climatically and geographically favorable for development of
complex and serious pest problems has created a large number
of unoccupied niches. As may be expected, rapid colonization
of this new crop by native fauna has occurred. Euryphagous
species, stenophagous species adapted to wild leguminous
species, and stenophagous species that appear to have shifted
their host preferences are involved (3). In addition to
these, annual immigrants, originating from both domestic and
foreign sources, are unable to overwinter successfully in
much of the South and are serious pests in the area. The vel-
vet bean caterpillar, *Anticarsia gemmatalis,* and soybean
looper, *Pseudoplusia includens,* are examples of the latter.

However, the high potential for insect damage to soybean
in the South did little initially to stimulate additional
research. An important reason for the slow start of research
on soybean insect pests in this area was the influence of
control measures being used for cotton insect pests. A high
percentage of soybean growers in the southern states were also
cotton growers. Conventional insecticides used on cotton were
cheap and effective, and it became common practice to apply
these same insecticides at the rates recommended for control
of cotton insect pests. Little thought was given to the
serious problem of environmental pollution, or even to
residues in the beans and oil, during the early stages of
expanded soybean production in the South. As recently as
1969, only 8 SMYs (scientific man years) were devoted to
research on insect pests of soybean in the United States.

Consideration of the following factors created the demand
for a substantial increase in research on soybean insect
pests—

1. increasing awareness that soybean in the southern
   states was at much greater hazard from attack by
   insects than had been experienced in areas of traditional production;

2. recognition that the magnitude of the problem resulting from the numbers and diversity of pest species

involved was such that an effective approach would
require a high degree of cooperation and coordination
of research effort;
3. indications that control of insect pests of soybean
   was in danger of following the same catastrophic path
   as that for cotton pests;
4. relatively low per acre value of the crop, making it
   economically impossible to tolerate high per-acre
   costs of insect control;
5. residues in the crop resulting from contamination by
   organochlorine insecticides applied directly to the
   crop, and soil residues resulting from applications to
   other crops rotated with soybean, or from drift of
   insecticides applied to other crops.

Responding to the obvious need, there was a substantial
expansion of research effort on soybean pests. By 1973, the
number of personnel fully involved in research on soybean
insects had increased to 30 SMYs. Ten state experiment stations and federal laboratories were involved.

DEVELOPMENT OF A STRATEGY FOR RESEARCH ON SOYBEAN INSECT PESTS

The problems of controlling insect pests of soybean were
recognized as being enormously complex and difficult. Soybean is attacked directly by many species. All parts of the
plant, aboveground and belowground, are attacked. In addition
to direct attack, more than two dozen species are known to
transmit soybean diseases, principally viruses. Ford and
Goodman (4) list eight species of major insect vectors involved in the transmission of ten virus diseases of soybean
known to occur in the United States. Their list did not include at least a dozen additional species known to transmit
soybean mosaic virus and three species of Pentatomidae that
are known vectors of the yeast-spot disease of soybean (5).

Of the 50, or more, species that attack the plant
directly, about 20 are considered to be major pests in the
South.

Consideration of these factors suggested development of a
regional research project as an approach to the problem. Such
a project was initiated as Regional Project S-74 "Biology and
Control of Arthropods on Soybeans." Activation of this
project was followed closely by organization of the NSF/EPA/
USDA "Integrated Pest Management Project" (the "Huffaker
Project") on "The Principles, Strategies, and Tactics of Pest
Population Regulation and Control in Major Crop Ecosystems."

Fortunately, many of the participants in both of these
projects had also had experience in cotton insect research.

Their experience with problems involved with cotton insects
convinced them that IPM was the only approach that would
prevent the development of similar problems with soybean
pests. Thus, it was relatively easy for participants to
adopt the IPM concept. The strategy developed on the soybean
subproject was one that could be expected to avoid the serious
mistakes of the past in insect control, or at the least hold
these mistakes to a minimum, while conducting needed research.
It was the conviction of the Soybean Subproject personnel that
this could best be done by making use of all available data to
formulate simple, interim IPM systems for use while data for
more sophisticated systems could be developed. Such systems
would—

1. take advantage of the soybean's ability to compensate
   for substantial injury without loss of yield or
   quality;
2. discontinue use of organochlorine insecticides for
   control of soybean insects;
3. use minimum rates of application of effective insecticides, choosing wherever possible the least environmentally disruptive and most highly selective, so that
   maximum effectiveness of native natural enemies could
   be realized;
4. base recommendations for treatment on the most
   accurately determined economic injury levels available;
5. treat only when necessary as determined by pest
   population assessment.

It was believed that adopting such simple systems would:
(1) maintain losses from soybean pests at economically acceptable levels until sufficient data could be collected for
developing more sophisticated IPM systems; (2) hold selective
pressure of insecticides to such low levels that development
of resistant populations would be unlikely to occur; (3) reduce substantially the burden of environmental pollution;
(4) conserve populations of native natural enemies; (5) reduce insecticide residues in the crop to tolerable levels or
eliminate them; and (6) circumvent the efforts of many
segments of the insecticide industry to develop systems based
exclusively on indiscriminate use of chemicals.

Adopting this strategy required that major, immediate
efforts be devoted to: (1) establishing minimum rates of
applications and the most effective schedule of timing
applications of insecticides for those species known to
require control; (2) developing more realistic, accurate
economic injury levels for these species; (3) improving
procedures for monitoring populations; and (4) assessing the

pest status of other species.

Concurrently, research emphasizing both basic and applied aspects, would concentrate on ecological studies of insect pests, entomophagous insects, and pathogens; economic injury, plant growth, and insect-disease relationships; and tactics of pest population regulation. New tactics would be integrated into the systems as rapidly as research made them available.

MAJOR ACCOMPLISHMENTS OF RECENT RESEARCH ON SOYBEAN INSECT PESTS

My discussion will emphasize two major themes: (1) progress made in developing effective, economical, environmentally acceptable, and dynamic IPM systems for insect and related pests of soybeans; and (2) identification of problems that require additional coordinated, highly cooperative, interdisciplinary research in the immediate future. The discussion is based on the research of a large number of personnel from several institutions and agencies, but mainly from researchers of the six universities involved in the NSF/EPA, IPM Soybean Subproject. However, I should like to emphasize and acknowledge the substantial contributions of personnel associated with Southern Regional Research Project S-74, state experiment stations, CSRS, ARS, and grower organizations.

If only one accomplishment could be considered most important for this research effort, it would be the NSF/EPA Integrated Pest Management Project's approach to agricultural research. No other approach yet devised could have resulted in the development in such a short time of effective, economical, ecologically acceptable, stable pest management systems for a major crop. It is particularly significant that this was done for an essentially new, 55 million-acre crop, 20 million of which is in an area that is at relatively high hazard from a large complex of important and potentially important pests. Many of these species had been studied relatively little before this project was initiated. Closely coordinated, cooperative efforts of a group of scientists from diverse disciplines were required to make such rapid progress on a complex and difficult problem. Such an effort could not have been developed by any other approach.

The strategy of beginning the project by putting together prototype pest management systems, crude and simple as they were, as structures around which the research effort was built, also contributed greatly to the success of the Soybean Subproject. Although relatively few data were available upon which to build such systems, they provided a nucleus around which more sophisticated systems could be built. Of equal

importance, their development effectively blocked vigorous
efforts being made to influence growers to adopt a system of
controlling soybean pests based on repetitive applications of
broad spectrum insecticides (6).

ECOLOGICAL STUDIES ON INSECT PESTS, ENTOMOPHAGOUS INSECTS,
AND PATHOGENS

Significant contributions have been made to the biology
and ecology of many members of the pest complex and their
natural enemies. Whitcomb et al. [in (7)] found that the
velvetbean caterpillar, *Anticarsia gemmatalis*, is capable of
overwintering in Florida at latitudes south of approximately
23°. They also identified four species of introduced plants
that serve as winter hosts. The finding that larvae appear
in the fields in southern Louisiana by or before the time
northern Florida is invaded suggests that immigrating moths
from areas outside the continental limits of the United States
are probably the most important sources of infestation each
year by this pest. If additional research should prove that
the overwintering population in southern Florida contributes
significantly to velvetbean caterpillar populations that occur
in areas to the north, research aimed at controlling the pest
during winter on a relatively small area of the Florida
peninsula might prove to be especially profitable. Direct
control of the pest or indirect control by destruction of its
overwintering hosts are attractive possibilities.
Another important pest, the soybean looper, *Pseudoplusia
includens*, has a similar pattern of overwintering. Like the
velvetbean caterpillar, it is not capable of surviving the
winter in major soybean-producing states. It does survive
in southern Florida and extreme southeastern Texas. However,
it appears that the major sources of infestation for most of
the southern states are areas of Latin America. Unlike the
velvetbean caterpillar, the soybean looper assumes importance
as a pest of soybean only in areas where cotton occurs in the
ecosystem. This phenomenon first attracted attention when it
was observed that outbreaks were occurring regularly in areas
of Louisiana where cotton and soybean were rotated but never
occurred in areas where soybean was grown as a monoculture on
recently cleared lands. Also, the looper was not a pest in
areas where soybean was grown in rotation with rice, or with
sugar cane and corn. Burleigh (8) suggested that excessive
use of insecticides for control of cotton pests affected
natural control agents adversely and thus was responsible for
irruptions of loopers on soybean acreage nearby. Further
study (9) showed that this was not the case. Instead, they

found that adult *P. includens* required a source of carbohydrate before females could produce the normal complement of eggs. Nectar secreted copiously by floral and extrafloral nectaries of cotton produced an abundance of carbohydrate. Moths that fed on nectar provided by cotton oviposited normally. This information is especially interesting for two reasons. Clearly, it shows that diversity does not always promote stability. It also suggests the probability that the causes of many insect outbreaks considered to be insecticide-induced may not be entirely understood.

Expansion of soybean acreage in the southern states provided the southern green stink bug, *Nezara viridula,* with an abundance of a favorable host, and changed the species from a relatively minor pest into one of major importance. This species is now considered to be the most important pest of soybean in many of the states bordering the Gulf of Mexico and Atlantic Ocean. Because it had been considered relatively unimportant, little research had been devoted to the species in the United States. Information on its biology and ecology was available from research on other crops in foreign countries, especially in Japan (10,11). Lack of information on the population dynamics of the species in the United States created a need for research on the pest. The most apparent need was for information on its spatial and temporal distribution in relation to soybean for use in developing an effective management system. This research was given high priority and an impressive amount of useful data has been accumulated on the ebb and flow of populations of *N. viridula* through space and time. It overwinters as a diapausing adult. The diapause is facultative and of varying intensity among individuals. Diapause development is completed by early February at the latest, during mild winters at more southerly latitudes of Gulf Coast states. Overwintered adults feed and breed on a variety of crops and wild hosts. Clovers, crucifers, and wild leguminous hosts are especially favored and effective in producing first generation adults, which feed on the same hosts. However, the southern green stink bug appears to be adapting rapidly to corn and wheat. Large populations are developing with increasing frequency on these two hosts. Both first generation and surviving overwintered adults are attracted to soybean. Early planted fields often support these insects in considerable numbers. Although oviposition is heavy on vegetative soybeans, the crop at this stage of development is a poor host. Populations are barely maintained on soybean prior to the beginning of pod filling.

Populations of the southern green stink bug require abundant food, which is provided only by the reproductive parts of its host plants, for rapid population growth. It is

this weak point in its biology, plus its highly developed
abilities to find hosts in the proper stage of reproductive
development, that can be exploited most effectively for population regulation.

Other species of stink bugs also attack soybean. The
most important are the green stink bug, *Acrosternum hilare*,
and the brown stink bug, *Euschistus servus*. In some soybean
ecosystems their combined populations may be nearly as large
as that of the southern green stink bug. Their damage potential is similar and for practical purposes the three species
may be considered as one.

A useful method for predicting population trends of the
southern green stink bug has been developed. It is based on
dissection of adults to determine their reproductive status
and the degree of parasitism by the tachinid, *Trichopoda
pennipes*, and on phenology of host plants. The method also
applies to the other species of stink bugs (6).

More detailed ecological information has made possible
both substantial savings in amounts of insecticide required
for control of most pests and in time and expense involved
in monitoring pest populations. Three examples are of particular interest: (1) stink bugs are unable to increase their
populations on soybean until pods begin to fill; (2) the soybean looper does not develop damaging populations except in
situations where cotton and corn are produced in rotation,
and cotton nectar or some other nectar-producing crop fur-
nishes the carbohydrate required for moths to produce eggs;
and (3) soybean attracts the corn earworm during the period
of flowering and early pod set only, and then only in fields
where growth of the plants has not been sufficient to cause
the canopy to close. Knowledge of these factors that affect
the temporal and spatial ebb and flow of pest populations
through the soybean ecosystem allow for population-monitoring
activities to be concentrated on that part of the crop that
is susceptible to attack. Varieties whose growth and development to susceptible stages are asynchronous with populations
of stink bugs and corn earworms may be safely ignored; the
soybean looper needs no consideration as a pest except where
it is grown in association with cotton.

Changes in Economic Injury Levels and Pest Status

The remarkable ability of soybean to compensate for
injury by insects, without loss of yield or adverse effect on
quality, substantially reduces the difficulty of developing
effective IPM systems for control of its pests. Until recently, need for application of insecticide was almost entirely

based on subjective estimates of the amount of visual damage being caused by pest attack. Too often, such "guesses" resulted in overtreatment.

Concentration of research toward the development of accurate, reliable economic injury thresholds for some of the most important pests of soybean has resulted in rapid progress, and has made possible substantial reductions in use of conventional chemical insecticides. Data obtained during the last five years have increased the pest population levels thought to cause economic injury from one bug per three feet of row to one per foot of row, and from one larva per three feet of row to three larvae per foot of row, for the southern green stink bug and corn earworm, respectively. Similarly, drastic changes have been made for other pest species.

When a new crop is planted in an area, there is usually a period of time during which it is colonized by various pests. This process is likely to continue until all vacant niches are occupied. This appears to be the case for soybean in the southern United States. Three new pests of the crop have attracted attention: a cerambycid stem borer, *Dectes texanus*; a platystomatid fly, *Rivellia quadrifasciata*, whose larvae attack nodules; and the tobacco budworm, *Heliothis virescens*, whose larvae feed on foliage and pods. *Rivellia* appears to have colonized soybean readily and successfully although it was not discovered to be a pest until 1975 (12). *Dectes* seems to be adapting to soybean more slowly but it is causing noticeable damage in restricted areas with increasing frequency. *H. virescens* is adapting to soybean, but more slowly than the other two. However, an obvious change has occurred during the last few years and this species may eventually develop to key pest status.

Natural Control Agents

A fundamental principle of IPM is that natural control agents should be used to regulate pest populations. The principle stresses making maximum use of both native and introduced species, and has been especially useful in developing IPM systems for soybean pests.

The diversity and complexity of insect pests in soybean ecosystems has been discussed above. However, as complex and involved as it is, the complement of insect pests is relatively simple compared to the natural control agents that play a major role in regulating pest populations. Many of the pest species are attacked by as many species of natural control agents as the total number of pest species that attack soybean. Effectiveness of these natural control agents is a key

to the progress that has been made until now in developing
effective, economical pest management systems for soybean
that growers have accepted and are using enthusiastically.

Native polyphagous predators, native and exotic parasites, and native and exotic microbial pathogens have been
evaluated. Their immediate and potential value as important
components of IPM for soybean has been confirmed. However,
vast amounts of additional research will be required to
discover ways of making maximum use of these indispensable
control agents.

Research on IPM systems shows conclusively that the
importance of native polyphagous predators has been underestimated. Effectiveness of polyphagous predators often has
been questioned because of their lack of host specificity.
Rather than being a weakness, lack of host specificity is the
major characteristic that makes a polyphagous predator complex
so effective. Much previous research on general predators has
involved evaluation of individual species rather than the
species complex. Considering only the impact of individual
species of predators inevitably leads to the wrong conclusion
that polyphagous predators are relatively ineffective.

The dictum that diversity builds stability is especially
applicable to use of polyphagous predators in IPM systems for
soybean. Both diversity of species and lack of host specificity of polyphagous predators provide this complex of
natural agents with the versatility to be effective in annual
row crops, where the more narrowly host-specific agents have
been conspicuously ineffective. The effectiveness of the
polyphagous predator complex for control of soybean pests is
effectively demonstrated by improper use of insecticides.
For example, one application of methyl parathion, improperly
timed, has been shown to decimate populations of the polyphagous predator complex so severely that corn earworm populations reduced yields in treated plots by 67% compared to
those in untreated plots (7). Application of the same
insecticide, properly timed, almost doubled the yield compared
with the untreated plots.

Although native parasitic species generally have not
proved as effective for regulating populations of pest species
as polyphagous predators, they are valuable components of the
natural enemy complex. In a few cases they provide spectac-
ularly effective control. In some areas populations of first
instar larvae of the green cloverworm, *Plathypena scabra*, are
controlled by more than 90% by one species, *Apanteles margin-
iventris* (7).

The potential for introduction and colonization of
exotic parasites has not yet received adequate attention in
developing IPM systems for soybean. That great potential

exists for such natural control agents is amply demonstrated by experience with *Pediobius foveolatus,* the Mexican bean beetle parasite. Excellent results have been obtained with this species both by the traditional method of introduction and colonization and by annual releases in areas where it cannot overwinter successfully. *P. foveolatus* has given spectacular control of *E. varivestis* in Florida[1]. Annual releases of the parasite into trap crops of lima bean, which effectively concentrate populations of overwintered adult *E. varivestis* into small areas, have been incorporated into the IPM system in North Carolina (7). This example provides ample reason for optimism that much potential exists for progress in rearing and releasing natural control agents periodically as well as in introducing and colonizing exotic species.

Recent research clearly shows that there is great unexploited potential for effective use of microbial pathogens in IPM systems. These agents, both native and exotic, have demonstrated value, and attractive possibilities exist for selective use to control several major pests.

Nomuraea rileyi, a naturally occurring entomogenous fungus with a wide host range among lepidopterous larvae, has proved to be effective for control of such major pests as the velvetbean caterpillar, soybean looper, and corn earworm. Sprenkel and Brooks (13) demonstrated an effective method for inducing epizootics of *N. rileyi* earlier in the season and at lower population densities than normally occurs. Their method consists of distributing 3 mm fragments of tobacco budworm larvae, *H. virescens,* killed by the fungus and stored at low temperature until needed, over field plots. The method provided effective inoculum for as long as three weeks in treated plots.

Another fungus, *Entomophthora gammae,* occurs each year where soybean looper populations reach high densities, in epizootics that usually provide control of 90% or more (14). Unlike *N. rileyi,* it is narrowly specific.

Two exotic nuclear polyhedrosis viruses applied at rates of 50 to 100 larval equivalents per acre give control of two major pests of soybean that compares favorably in effectiveness with recommended insecticides. One, the soybean looper NPV, was imported from Guatemala; the other, the velvetbean caterpillar NPV, was imported from Brazil. Laboratory, small field plot, and in the case of the soybean looper, large scale field experiments have demonstrated conclusively that both viruses have great potential for use in IPM systems. They are

[1]*F.W. Maxwell, personal communication*

narrowly specific, do not require such carefully timed
application as conventional insecticides, and have none of
the adverse environmental effects characteristic of insecticides.

In spite of all the obvious advantages of microbial
pathogens for use in IPM systems, it appears highly unlikely
that these agents will be available for use in the near
future. The question of possible human health hazards
connected with their use has not been resolved and resolution
of this question appears to be far from imminent. As a result of this unhappy situation, research and development of
these demonstrably useful agents has been substantially
reduced. In some cases it is at a standstill.

Selective Use of Insecticides

It is generally recognized that few truly narrow-
spectrum chemical pesticides, of the sort that would be
especially useful in IPM systems, have been discovered. However, it has been pointed out that pesticides having very
broad spectra of activity can be used selectively (15).
Substantial progress has been made by employing this tactic
in the strategy for developing IPM systems for soybean pests.
The lack of key pests in soybean ecosystems, plus the ability
of soybean to compensate for substantial amounts of injury
without adverse effects on yield, allows for the most selective of all ways of using insecticides—not to apply any.
The present situation permits this on a high percentage of
the total acreage in the United States, even in areas that
are at greatest hazard from insect attack. In addition,
substituting minimum rates of application of the less persistent organophosphorus and carbamate insecticides for the
organochlorines, previously used at rates of application as
high as those used for control of cotton insects, has
achieved a high degree of selectivity.

Application of the principle of minimum dosage rates and
basing treatment on more realistic and accurately determined
economic injury thresholds has made it possible to reduce the
amount of insecticides recommended for control of soybean
insects for 1976 by one-third to one-half that recommended
for 1973 (7). This change in the kind and amount of insecticides used in soybean pest management systems has (1) helped
to conserve populations of beneficial insects, (2) substantially reduced environmental pollution, and (3) reduced the
probabilities of resistance to insecticides developing in
soybean pests.

Trap Crops

The trap crop principle has been developed as an effective, ecologically safe, and economical tactic of IPM systems for bean leaf beetle, southern green stink bug, and Mexican bean beetle. It exploits behavioral characteristics of these three pests that cause them to concentrate in very small areas of the soybean crop, less than 5% of the total, that are much more attractive than the remainder of the planting. In the case of the bean leaf beetle, a trap crop consisting of small areas near favorable hibernation quarters, planted 10 days to 2 weeks before the main planting, will attract and hold a very high percentage of overwintering adults; the Mexican bean beetle is similarly attracted to small plantings of lima bean; and the southern green stink bug is trapped in areas planted to soybean varieties that mature about 1 to 2 weeks before the remainder of the crop.

Populations of the bean leaf beetle and southern green stink bug are effectively controlled in areas planted to trap crops by use of the appropriate insecticide. The Mexican bean beetle can be controlled in the trap crop by use of insecticide, or by release of the parasite, *Pediobius foveolatus*, into the area at the appropriate time. In either case, use of the trap crop principle is an important component of IPM systems for control of soybean pests. It has the following advantages: (1) economy, (2) minimum adverse effect on beneficial insects, (3) reduction in levels of environmental pollution, and (4) minimal selective pressure on pest populations, thereby delaying or preventing development of resistant populations.

CURRENT STATUS OF IPM SYSTEMS FOR SOYBEAN

The strategy adopted for development of IPM systems for soybean has proved to be eminently successful. Effective, economical, environmentally acceptable, and stable IPM systems for soybean ecosystems have been developed and released to cooperative state extension services for recommendation to growers. They are being readily accepted and rapidly adopted.

Components of the systems vary from one area to another, depending on variations in the pest complexes. The following are two examples of such IPM systems (7).

A system developed for control of soybean pests in Louisiana incorporates the following tactics—

1. using economic injury thresholds as the basis for insecticide use decisions;
2. scouting fields at regular intervals, but only during

periods of growth and development in which the crop
is at risk, to assess damage and monitor populations
of both pest species and their natural enemies;
3. taking maximum advantage of the regulatory effects on
pest populations of biological control agents, including complexes of polyphagous predators, egg,
larval, and adult parasites, and microbial pathogens;
4. using trap crops of small acreages of early planted,
early maturing Group V soybean varieties for control
of the bean leaf beetle and bean pod mottle mosaic
virus for which it is the only important vector, and
for control of stink bugs;
5. using minimum rates of application of selective
insecticides when economic injury levels are exceeded.
The IPM system for control of soybean pests in North
Carolina incorporates the following tactics—
1. using economic injury thresholds as the basis for
insecticide-use decisions;
2. scouting at regular intervals to assess pest populations and damage;
3. protecting and enhancing arthropod natural enemies of
insect pests;
4. selecting early maturing varieties to reduce damage
by lepidopterous larvae, especially the corn earworm;
5. selecting planting dates to render some varieties
unattractive to the corn earworm through canopy
enhancement and asynchronization of flowering with
peak moth flight;
6. planting all double-cropped soybeans (late-planted
soybeans following wheat) in narrow rows to promote
rapid canopy development for corn earworm management;
7. using trap crops of lima beans to attract overwintered populations of Mexican bean beetle, and releasing
the parasite, *Pediobius foveolatus,* into the trap
crop areas at the appropriate time for control
of the beetle;
8. using minimum rates of application of selective
insecticides when economic injury levels are exceeded.
These systems are relatively crude and unsophisticated
but they are sufficiently dynamic to permit the ready integration of new or refined tactics as rapidly as they become
available.

PROBLEMS THAT REQUIRE IMMEDIATE INTERDISCIPLINARY RESEARCH

Economic Injury Thresholds for Pest Complexes

One of the most disappointing and disturbing results of

research on the NSF/EPA Soybean Subproject is the failure to
make any significant progress toward solving the problem of
economic injury thresholds for pest complexes. This has
proved to be exceedingly complicated and difficult research.
The development of accurate, reliable economic injury thresholds for some of the more important individual species of the
large insect pest complex that affects soybean has been one
of the outstanding contributions of research on the Soybean
Subproject. Good, dependable economic thresholds have been
developed for the bean leaf beetle, *Cerotoma trifurcata;* the
corn earworm, *Heliothis zea;* the green cloverworm, *Plathypena
scabra;* the southern green stink bug, *Nezara viridula;* the
soybean looper, *Pseudoplusia includens;* and the velvetbean
caterpillar, *Anticarsia gemmatalis.* All of these are important pests, and all frequently occur in the same field at the
same time in soybean-producing areas of the Gulf Coast and
South Atlantic states. When growers ask what is the economic
threshold for a complex consisting of several of these species,
there is no answer available. Also, there is no information
available for the grower who asks for a recommendation when
his soybean fields are infested simultaneously with all the
components of the complex listed above, none of which has
reached the economic injury level, but all of which are at not
less than 25% of the economic injury level. Soybean growers
are asking such questions with increasing frequency and
persistency.

The complexities of such problems and the difficulties of
solution are formidable indeed. But the complexity and difficulty is increased by several orders of magnitude when interactions with pathogenic organisms and nematodes that are
serious pests of soybean are considered. Injury by these
pests may affect economic injury levels for insect pests. For
example, in relatively simple situations such as the six-com-
ponent complex of insect pests described above, the bean leaf
beetle, *C. trifurcata,* is intimately associated with plant
pathogens. It is the only efficient vector of bean pod mottle
mosaic virus, one of the most important virus diseases of soybean in the United States. In addition to destroying foliage,
it injures soybean pods in two ways. Prior to pod-filling and
early maturity stages of development, it may destroy pods in
the same way that they are destroyed by corn earworm larvae,
Heliothis zea. However, beginning with the late pod-filling
stage of maturity, the inner wall of the bean hull becomes so
tough that the beetles cannot chew through to the developing
seeds. At this stage of development, however, they destroy
the outer layers of the pod, thereby providing an avenue of
entrance for various pathogenic organisms. A high percentage
of pods injured in this manner are destroyed by invasion of

microorganisms.

There are additional elements of complexity provided by feeding of bean leaf beetle larvae on roots and nodules of soybean. In addition to providing avenues of entrance to pathogenic organisms, such injury at moderate to heavy levels of infestation reduces nitrogen fixation by as much as 75% to 90% during various stages of soybean growth and development. Peak levels of nodule attack are during the mid- to late-vegetative and early- to mid-pod-filling stages of development[2].

Other insect pests of soybean are also associated with plant pathogens in addition to the large number of species that are vectors of virus diseases such as tobacco ringspot (soybean bud blight) and soybean mosaic virus. Stink bugs are directly associated with yeast spot disease (5) and the three-cornered alfalfa hopper, *Spissistilus festinus,* with sclerotial blight (16), for example. These interactions have not thus far been considered in establishing economic injury levels for these pests.

Economic Injury Levels for Insects that Attack Belowground Parts of Soybean

It is a disconcerting fact that no economic injury level has been determined for any pest of the belowground parts of soybean. The importance of a healthy root system to plant growth is well recognized. The role of root nodules in fixation of atmospheric nitrogen adds another dimension to the importance of belowground portions of soybean. Yet, very little research has been devoted to the insect pests that attack soybean roots and nodules. Apparently none has been devoted to the determination of economic injury thresholds for species capable of causing damage to the roots and nodules. The lack of research is emphasized by the discovery during 1975 that the larva of a platystomatid fly, *Rivellia quadrifasciata* Macquart, was destroying as much as 40% of the nodules of soybean in some fields in Louisiana (12). The species had not been recognized previously as a pest of soybean or any other crop. Undoubtedly, its injury has been confused with that caused by larvae of the bean leaf beetle, as the two species produce similar kinds of damage. The possibility of other belowground pests of soybean going unrecognized is disturbing.

[2]*L.D. Newsom, unpublished data*

Interdisciplinary Conflicts of Interest Related to Pest
Complexes

Until recently, growers have been asking questions mainly
related to economic injury levels involving insect pest complexes. They have now begun to be concerned about the effects
on yields of injury caused by complexes of insects, viruses,
fungi, bacteria, and nematodes. The interactions involved in
these complexes are enormously complicated and difficult to
study. The effects may be additive, synergistic, or anta-
gonistic. However, as Powell (17) cogently pointed out in
his discussion of the interaction between nematodes and fungus
diseases, none of the entities involved can be detached from
the complex when the effects of other members of the biotic
community are being considered.

Some examples that illustrate the critical need for
interdisciplinary research to understand the complex interactions among the host of organisms that affect soybean have
developed during the last five years. Soybean foliage, stems,
and pods are infected by dozens of fungal, bacterial, and
viral diseases (18-21). All parts of the plant are affected.
Generally, pathologists have not concerned themselves with
research aimed at determining economic injury levels for any
of these disease-producing agents. Probably major dependence
on host-plant resistance and cultural practices for control
of most pathogens of field crops has been responsible for lack
of interest in economic injury levels. Both measures, when
used, are involved automatically by or before the time of
planting.

Until recently use of fungicides for control of plant
pathogens of soybean was impractical. Available, effective
fungicides were either nonexistent or much too expensive for
a crop of such low per acre value as soybean. However, with
the development of some of the newer fungicides, notably
benomyl, the situation changed drastically. One or two properly timed applications of benomyl at 0.5 to 1.0 pound per acre
often result in substantial increases in yield. Increases av-
eraging 4 to 5 bushels per acre are common especially in areas
characterized by relatively high rainfall during summer and
early fall. Tremendous interest and enthusiasm has been generated among growers for treatment of soybean with benomyl during the period of early pod development. Approximately one-
fourth of the acreage in Louisiana was treated during 1976, and
acreage treated during 1977 will increase substantially. Thus
far, pathologists do not completely understand what is responsible for striking increases in yields in some localities, for
some varieties and during some years, when no increases occur
in other situations. There is obvious control of several of
the foliage, stem, and pod diseases, especially pod and stem

blight, *Diaporthe phaseolorum* (Cke. & Ell.) Sacc. var. *sojae* (Lehman) Wehm.; anthracnose, *Colletotrichum truncatum* (Schw.) Andres and W.D. Moore; and target spot, *Corynespora cassicola* (Berk. and Curt.) Wei.

Unfortunately, no information is available for growers to use in making decisions on whether or not their crop should be treated with a fungicide. Even more unfortunately, it appears that little, if indeed any, research is underway to develop the necessary information for making rational decisions about use of foliar fungicides on soybean. It is generally observed that treated plants retain their foliage longer than untreated with a consequent delay in maturity of about 1 week to 10 days. The situation is even more complicated, because benomyl has an adverse effect on the entomogenous fungi, *Nomuraea rileyi* and *Entomophthora gammae*. These pathogens are of great value in control of lepidopterous defoliators of soybean. Applications of benomyl for control of fungal pathogens of soybean are made at a time when adverse effects on natural control agents of lepidopterous pests are most critical.

Nematodes have long been recognized as major pests of soybean. Species of six genera have been reported to be pathogenic to soybean (22). Among these the soybean cyst nematode, *Heterodera glycines* Ichinoche and root-knot nematodes, *Meloidogyne* spp., are devastating pests in some areas. However, no research has been done on economic injury thresholds for these two major pests. The pest status of other species reported to attack soybean has not been determined. Although considerable research has been done on nematode-fungus interactions, none appears to have been done on nematode-insect interactions. Conclusions made by Taylor and Wyllie (23) that additional research is required for understanding the interrelationships between root-knot nematodes and *Rhizoctonia* on soybean apply even more strongly to interactions involving insects, nematodes, and fungi.

A conflict of interest exists between entomologists and nematologists in situations where preplant soil applications are made for nematode control, using nematicidal chemicals that also have high insecticidal action. Some of these chemicals depress populations of native natural enemies of insect pests to the extent that populations of pest species, *Heliothis* spp. for example, are released and may increase dramatically. Applications of these broad spectrum systemic chemicals to large acreages of soybean may have seriously detrimental effects on populations of native natural enemies. Undesirable effects may be both direct and indirect. Direct effects are most likely to occur on species of hemipterans such as *Geocoris, Spanogonicus, Nabis,* and *Orius* that feed on both plants and insects. Indirect effects resulting from destruction of major prey species such as thrips, spider mites, and leafhoppers are probably more important than direct effects. They are of two kinds: (1) direct toxic effects, and (2) secondary poisoning of predators causing them to leave treated fields or starve.

Modest, interdisciplinary research efforts are underway to evaluate these complex problems. Entomologists and plant pathologists appear to be more actively involved in study of the adverse effects of benomyl on entomogenous fungi than are entomologists and nematologists on interactions involving use of nematicidal-insecticidal chemicals on soybean. There is need for much more vigorous efforts on both problems. There are numerous additional interactions similar to these in soybean ecosystems. Their complexity is such that they cannot be effectively studied except by well-coordinated, cooperative interdisciplinary research of the sort exemplified in the NSF/EPA IPM Project. Clearly, the approach taken on this unique project is the most effective way yet devised for doing research on complex interactions involving a variety of organisms. It appears appropriate that it be expanded to include more disciplines at maximum levels of participation.

There is an urgent need for development at the national level of some mechanism for continuing and expanding the sort of coordinated cooperative, interdisciplinary effort that has proved to be so effective in the NSF/EPA IPM Project.

REFERENCES

1. United States Department of Agriculture. 1977. Prospective plantings. Crop. Rep. Board, Stat. Rep. Serv., April 14. Washington, D.C.
2. ———. 1950-77. Crop production. Annual Summaries. Crop Rep. Board, Stat. Rep. Serv., Washington, D.C.
3. Turnipseed, S.G. and M. Kogan. 1976. Soybean entomology. Annu. Rev. Entomol. 21:247-82.
4. Ford, R.E. and R.M. Goodman. 1976. Epidemiology of soybean viruses. pp. 501-12 *in* World Soybean Research. L.D. Hill, ed. The Interstate Printers and Publishers, Danville, Illinois.
5. Daugherty, D.M. 1967. Pentatomidae as vectors of yeast spot disease of soybean. J. Econ. Entomol. 60:147-52.
6. National Science Foundation. 1975. Integrated pest management: the principles, strategies, and tactics of pest population regulation and control in major crop ecosystems. Prog. Rep., Vol. 1.
7. ———. 1976. Integrated pest management: the principles, strategies, and tactics of pest population regulation and

population regulation and control in major crop ecosystems. Prog. Rep., Vol. 1.

8. Burleigh, J.G. 1972. Population dynamics and biotic controls of the soybean looper in Louisiana. Environ. Entomol. 1:290-94.

9. Jensen, R.L., L.D. Newsom, J. Gibbens. 1974. The soybean looper; effects of adult nutrition on oviposition, mating frequency and longevity. J. Econ. Entomol. 67:467-70.

10. Kiritani, K. 1963. Oviposition habit and effect of parental age upon the post-embryonic development of the southern green stink bug, *Nezara viridula* L. Jap. J. Zool. 13:88-96.

11. Kiritani, K. and T. Sasaba. 1969. The differences in bio- and ecological characteristics between neighboring populations in the southern green stink bug, *Nezara viridula* L. Jap. J. Ecol. 19:177-84.

12. Eastman, C.E. 1976. Infestation of root nodules of soybean by larvae of the bean leaf beetle, *Cerotoma trifurcata* (Forster) and the platystomatid fly, *Rivellia quadrifasciata* (Macquart). Ph.D. Dissertation, Louisiana State University.

13. Sprenkel, R.K. and W.M. Brooks. 1975. Artificial dissemination of *Nomuraea rileyi*, an entomogenous fungus of lepidopterous pests on soybeans. J. Econ. Entomol. 68:847-51.

14. Brousseau, D.E. 1975. A field study of the seasonal history of *Entomophthora gammae* Weiser and its relationship to the soybean looper, *Pseudoplusia includens* (Walker). M.S. Thesis, Louisiana State University.

15. Newsom, L.D., R.F. Smith, W.H. Whitcomb. 1976. Selective pesticides and selective use of pesticides. pp. 565-91 *in* Theory and Practice of Biological Control. C.B. Huffaker and P.S. Messenger, eds. Academic Press, New York.

16. Herzog, D.C., J.W. Thomas, R.L. Jensen, L.D. Newsom. 1975. Association of sclerotial blight with *Spissistilus festinus* girdling injury on soybean. Environ. Entomol. 4:986-88.

17. Powell, N.T. 1963. The role of plant-parasitic nematodes in fungus disease. Phytopathology 53:28-35.

18. Athow, K.L. 1973. Fungal diseases. pp. 459-89 *in* Soybeans: Improvement, Production, and Uses. B.E. Caldwell, ed. American Society of Agronomy, Madison, Wisconsin.

19. Kennedy, B.W. and H. Tachibana. 1973. Bacterial diseases. pp. 491-504 *in* Soybeans: Improvement, Production, and Uses. B.E. Caldwell, ed. American Society of Agronomy, Madison, Wisconsin.

20. Dunleavy, J.M. 1973. Viral diseases. pp. 505-26 *in* Soybeans: Improvement, Production, and Uses. B.E. Cald-

well, ed. American Society of Agronomy, Madison, Wisconsin.

21. Sinclair, J.B. 1976. Seed-borne bacteria and fungi in soybeans and their control. pp. 470-78 *in* World Soybean Research. L.D. Hill, ed. The Interstate Printers and Publishers, Danville, Illinois.
22. Edwards, D.I. 1976. Nematodes and nematode-fungus interactions on soybeans. pp. 629-33 *in* World Soybean Research. L.D. Hill, ed. The Interstate Printers and Publishers, Danville, Illinois.
23. Taylor, D.P. and T.D. Wyllie. 1959. Interrelationship of rootknot nematodes and *Rhizoctonia solani* on soybean emergence. Phytopathology 49:552. (Abstr.)

DISCUSSION

W. LOCKERETZ: You stated that application of methyl parathion causes decreased yields, by damaging natural enemies of pests. How often do you think this happens in farmers' fields without their knowledge?

L.D. NEWSOM: It does happen, but not at alarming rates in soybeans, because less than 20% of the acreage in the "high hazard" area in the South is treated in any one year. We've been very successful in holding off the indiscriminate use of pesticides on soybeans, and we hope we can continue to do so.

UNIDENTIFIED: How can you convince the grower that 40% defoliation won't affect yield?

L.D. NEWSOM: With great difficulty. But we have been able to do this surprisingly well, up to now. Many of these growers also grew cotton, and they know what can happen if you overuse chemicals.

UNIDENTIFIED: Are soybeans a reasonable crop to grow as a commercial venture in the southern areas? Does it make sense to encourage farmers to grow soybeans?

L.D. NEWSOM: It sure does. People make lots of money at it, and it is a very important crop.

UNIDENTIFIED: Plant tolerance is related to phenology—is that taken into account in setting thresholds?

L.D. NEWSOM: Yes. For example, we can use early-maturing varieties to get away from corn earworm damage, but there are problems with using these varieties.

UNIDENTIFIED: How do you view the future of tolerant soybean varieties?

L.D. NEWSOM: I have nothing but optimism for these varieties. They are potentially very important, but we have not yet

exploited them enough.
UNIDENTIFIED: How large a yield is needed in a tolerant variety for it to be accepted?
L.D. NEWSOM: You have to come very close to the yield of
current varieties to convince our growers to accept them.

APPLICATION OF COMPUTER TECHNOLOGY
TO PEST MANAGEMENT[1]

Dean L. Haynes
R.L. Tummala

Department of Entomology
Michigan State University
East Lansing, Michigan

INTRODUCTION

The numerous applications of computer technology make an impressive array; they include the majority of our society's economic activities, technological improvements, and communication systems. The exponential growth of computer technology seems to be driven by newly emerging economic needs and the ability of this technology to redesign itself. The unit price for a given level of computer output is continually decreasing. The needs of agriculture and more specifically pest management have not been an important factor in bringing on the age of computers. In fact, we have been fairly primitive users, and the conservative nature of our discipline has prevented us from keeping pace with other elements of our society.

To demonstrate the future role of computers in pest management, we will review their role in the past, and also discuss our program that relies heavily on computer technology. It is important to note that pest management includes several factors usually associated with environmental management, as efficient pest management involves many of the

[1]Michigan State University Agricultural Experiment Station Journal Article No. 8275. This work was supported by the Environmental Protection Agency Grant R80-3785, "Utilization of Pest Ecosystem Models in Pest Management Programs" and a Pilot Pest Management Program supported by Extension Service, USDA entitled On-Line Management of Field and Vegetable Crop Pests (1975).

complex interactions that characterize man's activity in his
biotic and abiotic environment. Research in pest management
using computer technology has tremendous potential to lead
the way into the future, when environmental problems will be
considered primary concerns rather than just products of man's
activity.

An approach to pest management that relies heavily on
computer technology is being developed, implemented, and refined by scientists at Michigan State University. This
coordinated research and development program was initiated in
1969 and has relied on the cooperative interactions of scientists in disciplines from agriculture, engineering, and
natural science, with endorsement and support from the Center
for Environmental Quality, the Agricultural Experiment Station, the Agricultural Extension Service, the Division of
Engineering Research at Michigan State University, the National Science Foundation, United States Department of Agriculture, and Environmental Protection Agency.

MULTIFACTOR CONTROL OF INSECT PESTS

Pest Management

There are many alternatives to pesticides as pest management tools. These include the development of plant resistance; control by parasites, predators, and competitors;
microbial controls; viruses; cultural controls such as crop
rotation, planting schemes (temporal and spatial) and fertilization; chemical attractants and repellents; feeding deter-
rents; and sterilization. Individual control programs have
evolved over the years into the classical approach to pest
management schematically depicted in figure 1. With this
approach, the particulars of an existing or potential pest
problem are first identified and then resources are directed
into discipline-oriented research programs to screen chemical
compounds for toxicity, prepare spray calendars, develop resistant host varieties, release parasites or predators, etc.
This classical approach is clearly a multifactor program in
pest control, and contains many of the features of a pest
management program. However, each technique is researched as
if it were, or must be, the total and ultimate solution to the
particular pest problem. Extension workers recommend the
technique that shows the greatest promise for achieving the
control. If a particular technique fails, the alternatives
are reevaluated and the research is continued. If secondary
pests develop, a new program of research is initiated. The
end result for any given pest situation is a set of

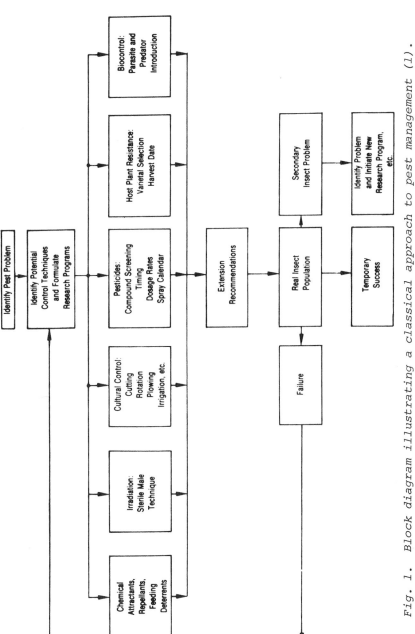

Fig. 1. Block diagram illustrating a classical approach to pest management (1).

recommendations with little or no flexibility to account for stochastic and regional variation. No provision is made for integrating the various techniques of control into more comprehensive strategies based on the ecological character- istics of the total ecosystem. Evaluation of the alternative control techniques often ignores the dynamic interactions be- tween the pest and its environment. The static recommenda- tions that are specified, in addition to lacking flexibility, do not consider the age structures and time synchronies of the interacting populations or the long-term economical and ecological constraints on the ecosystem. For many pests, the practical limits of such solutions have been reached. By its very nature, the classical approach to pest management research cannot fully exploit the potential of integrated strategies of control for improving reliability, as well as ecological and economical effectiveness.

Systems Models and Pest Management

In order to study the complex pest ecosystem inter- actions, it is necessary to use the well-developed systems techniques of the physical sciences. This is clearly demon- strated by the extensive systems modeling efforts taking place on economic pests of various crop ecosystems throughout the world (2-7). However, much of the information is used in a historical context, after the fact, rather than in an antic- ipatory context. Although the models do provide predictive capability, it is recognized that models are just simple abstractions that include only the dominant features of the pest ecosystems being studied. This, along with stochastic variation of abiotic parameters affecting the system, explains the deviation of long-term model predictions from actual observations.

On-Line Pest Management

In order to overcome these limitations, a new approach was described by Haynes et al. (8). This approach is referred to as "real time" or "on-line" pest management system to em- phasize that it is—
   1. based on models that integrate biological and envi- ronmental factors with chemical treatments and cultural practices; and
   2. continuously reviewed in light of changing meteoro- logical conditions, the ecological state of the eco- system, and the relative effectiveness of previous control strategies—hence the designation "on line."

Through this approach to pest management, as depicted in figure 2, multifactor control strategies can be systematically developed and modified from region to region according to day-to-day changes in weather, field, and economic factors. This on-line pest management system has the following essential operating features. Environmental data are obtained at selected sites around the state or region of interest. Summaries of this information are sent out immediately, via communication channel A, to appropriate users for insertion into their predictive pest management models. This information together with the biological data obtained from communication channel C serves as the basis for synchronizing these models with the actual pest ecosystem and for refining parameter estimates. The models, which incorporate the dynamics of the interacting

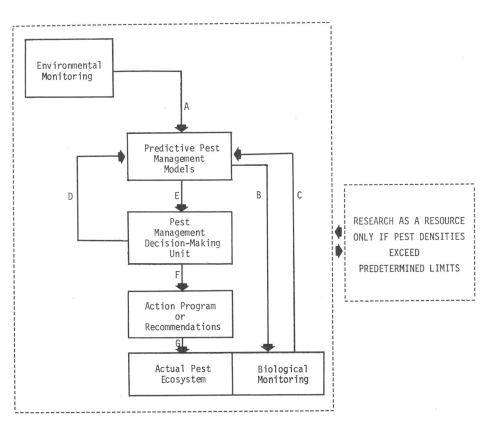

Fig. 2. Block diagram illustrating the basic components of a fully operational on-line pest management system (1).

populations, the quantitative features of the various management strategies, and the economical and ecological constraints, are then used to identify both the control strategy
alternatives and the points in time when critical biological
measurements (communication channel B) are to be made on the
pest ecosystem. These measurements characterize the state of
the ecosystem and are used to update the models periodically
and to monitor the effectiveness of the previous control
strategies. Agricultural extension personnel in the pest
management decision-making unit interrogate the management
models (communication channel D) and then, on the basis of
the output (communication channel E), alert growers (communication channel F) to the various control strategy options.
This system must function so as to guarantee that recommendations can be quickly transferred to growers so that appropriate action programs can be implemented. The speed, volume,
and complexity of the information flow clearly necessitates
the use of a computer-based communication system.

Models in On-Line Pest Management Programs

The ultimate objective in managing any pest ecosystem is
to maintain it, through selected controls, in such a state
that the interactions of man, pests, and the environment are
ecologically compatible and economically and socially feasible. This goal can be achieved through the development and
implementation of a series of dynamic models, which use on-
line measurements of selected biotic and abiotic features of
the pest ecosystem to predict the effect of various control
strategies. These models represent the keystone of an on-
line pest management program. In general, the type of model
we need will have high short-term predictability (three days
or less) but considerably lower long-term predictability due
to the stochastic nature of the microclimate. Clearly a major
need, which requires computer technology, is the periodic up-
dates of these models with current climatological and biological information. The principal task of such a computer
system is—
  1. to use real-time climatological and biological infor-
     mation to describe the current state of the pest eco-
     system and update the computer-based predictive pest
     management models;
  2. to interrogate a variety of models in order to deter-
     mine the acceptable near-term control strategy options;
  3. to communicate these options to users in such a manner
     that the control measures can be properly implemented.

Biological Monitoring in On-Line Pest Management Programs

The problem of updating models with current biological information is not simple. If a trained entomologist can look at a potential problem frequently during a season, he can evaluate population buildup of the pest and parasites, plant growth response, and progression of damage. On the basis of these observations he can make highly accurate predictions. In any survey of pest numbers, the timing, the distribution of ages in the population, the width of the age classes, and the distribution of mortality within age classes all affect the proportion of individuals counted in the population (9). If enough is known about these pest-specific factors, the proportion counted can be determined and the total season population level estimated. The time lag inherent in processing and interpreting biological information at present greatly limits our ability to implement an on-line pest management system.

One major deficiency with this method of data collection and analysis is the long time delay—typically six months to a year—from the time data are collected to the time results of the reduced data are inserted into the models. This delay makes it impossible to use this information to characterize the current state of the pest-crop ecosystem. In most cases, the analyzed data cannot be used in the decision process until the next growing season. Furthermore, because of these lengthy delays, existing techniques for investigating the dynamic behavior of the pest-crop ecosystem cannot be used in the eventual "on-line" control of the ecosystem. The other major deficiency with this method of conducting the biotic survey is the uncertainty surrounding the validity of the data.

The time lag inherent in a historic approach to biological monitoring renders most pest surveys useless to a pest management system, as depicted in figure 2, at least for online multifactor pest control within a given growing season at either the state or regional level. Several new approaches to the biological survey such as scouts, pest management field assistants, system users, or combinations of these approaches hold promise for overcoming these deficiencies. Each approach involves the application of sophisticated computer systems.

Other possible ways to improve the above method would be (1) to speed up manual data collection and analysis procedures, and (2) to reduce the data needed through cross-checking and correlation studies, thereby increasing the accuracy of the analyzed data. These improvements in the manual collection methods might be accomplished by defining a new and expanded role for agricultural extension.

Another approach is possible by automating certain

operations with the aid of the computer. Using present computer technology portions of a survey could be automated. For example, field samples could be collected manually or remotely and then analyzed with automated instruments that estimate leaf damage and crop development, or determine the age distribution of larval pest and predator populations and the parasitization rate. Output from these instruments would be coded electrical impulses, stored on paper tape, magnetic tape, or other storage medium, and then automatically inserted into a digital computer for analysis and data reduction. This approach would speed up both the time required to analyze the field samples and the time required for data reduction. The biggest advantage, however, would be to evaluate errors associated with the data independent of human activities.

Implementation of on-line pest management systems as depicted in figure 2 causes certain major changes in the method of conducting biological surveys. Certainly, the biological survey will always be important in confirming the basic reliability of the predictive pest ecosystem models and providing information to refine and optimize the models. There is no doubt, however, that it would be extremely advantageous to minimize the amount of data required in the biological survey. This can be accomplished by recognizing that the pest ecosystem unfolds during a given year according to physiological time. Hence, it is possible to take accurately timed and ultimately interpreted biological observations with a population model, only if the data can be placed on a time-temperature scale. Often the first biological event (the first insect to emerge, the first parasite to attack, etc.) cannot be observed. A method is needed to estimate and/or calibrate the observation in order to use the model effectively. In figure 3, a density observation taken at time t_1 will be identical to an estimate taken at t_2 but will have a vastly different meaning in terms of a model predicting plant damage. At t_1 most of the insect damage has not yet occurred, and at t_2 damage is well past the peak. When the estimates are taken from separate fields at two different times, there is no accurate way to assess the ultimate outcome in terms of crop damage caused by the pest population without the aid of a synchronized population model.

By using models and computer analyses we can shift the emphasis from making numerous biological observations throughout the year to the requirement that real-time microclimatic information must be known. This is precisely the approach being taken at Michigan State University. The predictive pest ecosystem models are being structured to accept climatological data as the principal real-time input. This information is used to synchronize the models with the actual pest ecosystems.

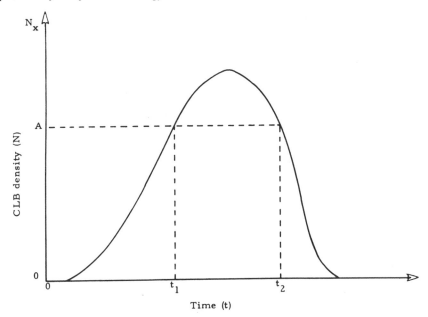

Fig. 3. Hypothetical total population incidence curve for an insect, where $N(t_1) = N(t_2)$ [see (1)].

These models are available as useful tools to the extension specialist for decision making. They can also inform him of the key time(s) during the year when certain biological samples should be collected and analyzed. Hence, heavy reliance need not, in general, be placed on receiving the results of the biological survey in "real time" since the models have high short-term predictability. But how is real-time climatological data obtained?

Environmental Monitoring for On-Line Pest Management Systems

Historically, agricultural scientists have relied on a national climatic summary for weather data. This monthly summary is typically available three months late, while the pest management models obviously require real-time data. Consequently, a weather-monitoring network must be developed that is capable of providing this critical information. For the past several years, researchers at Michigan State University have been attempting to develop environmental monitoring systems for pest management programs, both in a research and extension mode. These monitoring systems have now reached a stage where they can be incorporated into ongoing research and pest management programs.

The simplest and least expensive method of obtaining real-time remote-site environmental data would be to tie computer-based acquisition systems into one or more of the existing national weather service networks. One of the most promising of the services is the U.S. Department of Commerce's "Aviation Weather Network" (AWN) since it provides hourly weather summaries for hundreds of selected locations around the country and its format is computer compatible. In the past, this network has been used almost exclusively for air-traffic control and weather forecasting; thus, the environmental monitoring stations are located at airports. This source of climatic information was identified by Haynes et al. (8); however, the data disseminated to users is not exactly the type required by agricultural scientists. Data transmitted over this network include ceiling visibility, obstruction to vision, sea-level pressure, temperature, dew point, wind velocity, altimeter setting, and runway visual range. Only four of these nine parameters are of general interest to agricultural scientists. This contrasts sharply with the types of environmental data typically required (Table 1).

Utilizing this source of climatic information for an on-line pest management program requires a mini-processor (minicomputer) for selectively filtering the environmental information from the National Aviation Weather Network on an hourly basis. The useful information from this network for agricultural systems will be temperature, dew point, wind speed and direction. Figure 4 shows the location of the stations in the Lake States area that have been used to predict cereal leaf beetle developmental rates. This information will continue to be used to generate developmental maps for various pests considered within the on-line pest management program at Michigan State University and will be used to produce developmental maps for numerous pests similar to the cereal leaf beetle map shown in figure 5.

A second system is also now available, which provides daily input on temperature and rainfall from 37 locations in Michigan (fig. 6). This system is the Michigan Agricultural Weather Station Network, and provides on-line contact with climatic information with a 24-hour delay. This information is from agricultural areas and has a lower temporal resolution than AWN, but it has a higher predictive value than AWN. The computer system keeps an accumulated record of all this information throughout the growing season, and this weather file can be accessed from any remote terminal by a telephone.

The most recent development in environmental data collection at Michigan State University involves a satellite relay

Table 1. Representative Environmental Data Required for Computer Based On-Line Pest Management Programs (1)

Parameter	Number of Measurements Required per Site	Range	Accuracy	Threshold	Maximum Rate of Change of the Parameter per Unit of Time
Air temperature	1	-55-120°F	±2°F	—	2°F/min
Soil temperature	4(0,1,3,5 in.)	28-130°F	±2°F	—	2°F/min (surface) 2°F/hr (subsurface)
Wind speed	1	0-50 mph	±1 mph	1 mph	30 mph/sec
Wind direction	1	0-360°	±10°	1 mph	180°/min
Barometric pressure	1	27.0-31.5 in.Hg	±0.2 in.Hg	27 in. Hg	.33 in. Hg/hr
Humidity	1	0-95%	±1%	5%	20%/hr
Precipitation	1	0-4 in. rainfall	0.1 in.	0.1 in.	15-20 in./hr
Soil moisture	1	0-100% field capacity	±17%	0%	33%/hr
Light intensity					
Low	1	0-300 cal/cm^2/d	±5 cal/cm^2/d	0	—
High	1	0-10^4 ft/candles	20%	—	—

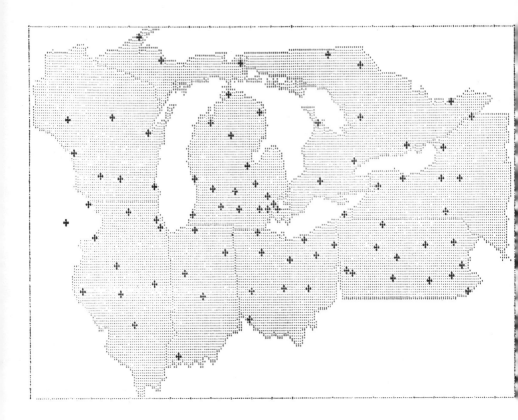

Fig. 4. Locations of national aviation weather network stations being recieved.

system. The system has been completely described in a final
report on contract NAS2-8773 to the National Aeronautics and
Space Administration on "The Implementation of a Data Collection and Satellite Relay System for Agricultural Pest Management." The work is a result of a two-year study, summarized
in a 150 page report entitled "The Evaluation of Alternative
Control Systems for Agricultural Pest Management." This
interim report to NASA (contract NAS2-8407) looks at different
environmental information systems to be used in conjunction
with cereal leaf beetle models in an on-line pest management
mode. At present, there are three transmitting platforms,
with antennas and sensors at Michigan State University. The
needs of on-line pest management require that we develop a
unified environmental information system that will receive and
store all information transmitted to it so that the information
can be utilized by extension specialists and other users
implementing pest management decisions.

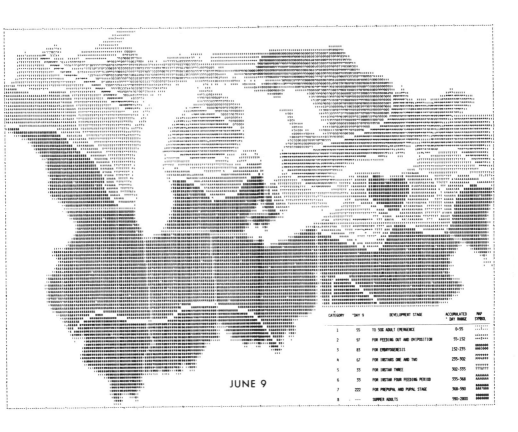

Fig. 5. Developmental map for the cereal leaf beetle
obtained from population models and environmental information
from stations in figure 4 for the day of June 9.

Operation and Maintenance of an On-Line Pest Management System

The emphasis of both basic and applied research on pest
management is rapidly moving toward multiple control strategies for pest regulation. This is distinctly different from
the recent past philosophy, which adopted the eradication concept on a local level using broad-spectrum pesticides. There
appears to be a tendency, though weak, to foster a convergence
of discipline-oriented pest control recommendations from weed
science, plant pathology, nematology, and entomology. The
need to integrate these plant protection recommendations at
some level above the grower seems obvious. However, the complexity of reaching such a goal is now just emerging.
The development of research models that represent dynamic

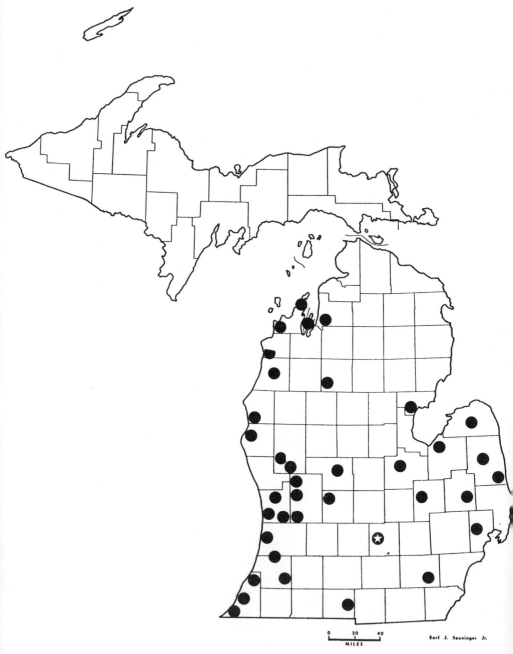

Fig. 6 *Locations of Michigan's agricultural weather station network.*

interaction within the pest ecosystem, driven and synchronized
by abiotic factors, requires a new look at agricultural extension delivery systems. Pest management models are of little
use unless they can be translated into an effective action
program. Agricultural extension plays a major role in on-line
management programs. The on-line system described in figure 3
is essentially an extension system utilizing a research component to assist in setting up the system.

On-line pest management requires a new and expanded role
for extension, one in which computer-based communication
systems are a standard element. A new attitude and philosophy
will be needed on the part of the extension field staff, because they will be the principal users of the on-line pest
management system and the principal source of current pest
data for updating models. This change clearly requires the
use of a high speed, two-way computer-based communication network.

Such a computer-based system has been developed at Michigan State University (10). At present the system can be
divided into the following communication, biological, and
environmental components:

Communications Features

CONTACT is a program designed to send messages among
pest management system users. Information of
all types can be sent in any format.

ALERTS is a program designed to allow extension
specialists to send pest alerts and control
strategy information to county extension field
staff and pest management field assistants.

AUTOMATIC SCHEDULING is a feature of the system
that allows users to receive periodic information about specific crops and pests without
specifically requesting it each time.

GETMSG is a program that allows a user to obtain
alerts and obtain messages through another
user's account.

MANUAL is a system that arranges for a complete set
of pest management computer system instructions
and program descriptions to be delivered to a
user.

DIRECTORY is a program that lists the addresses,
telephone numbers, and responsibilities of all
or specific individuals affiliated with On-Line
Pest Management Programs.

NEWS is a program that provides printouts of all
alerts and messages concerning field crops,
vegetables, or fruit crops.

SUGGEST is a feature of the system that allows users

to make suggestions or complaints about the
pest management computer system to the campus
staff.

CODEAPHONE is a program that provides listings of
telephone numbers of all vegetable, field crop,
and fruit code-a-phones.

INDEX is a feature that allows users to obtain an
up-to-date listing of all components of Pest
Management Computer Communication System.

HELP is the term employed by a user when he does
not understand some aspect of the pest management computer communication system. Appropriate information is automatically provided to
the user so that the problem can be solved
without direct contact with On-Line Pest Management staff.

Biological Features

Pest Status Summaries

(ALFALFA, SUGARBEET, ASPARAGUS, ONION, and POTATO)
are the programs that provide pest status summaries for these crops. Extension specialists
are recognized and allowed to issue alerts.
Information can be summarized on a county,
regional, or state basis. Data for the summaries are obtained from the monitoring
activities of the pest management field assistants, and other participants. Table 2 is a
typical summary of sugar beet pests as obtained
from a terminal in a county agent's office,
showing current pests of this crop for the week
ending August 7, 1976.

BLACKLIGHT is a program that provides county and
state summaries of insect information obtained
through the Michigan State University blacklight trap program.

BLITESUM is a program that provides county and state
summaries of late blight development information generated by BLITECAST.

Predictive Programs

BLITECAST is a late blight forecasting model for
timing of fungicide applications in potato
production. A program adapted for Pennsylvania
State University.

WEEVILCOST is a model designed to integrate the economics of alfalfa production with cutting
times, insecticide applications, and management
of alfalfa weevils.

POTATOPEST is a pest-crop ecosystem model for

simulation of potato losses caused by root-
lesion nematodes and generalized insect
defoliation.

Environmental Features

FORECAST is a program that provides local daily
weather forecasts and agricultural advisories,
including four-day temperature predictions.

DEGREEDAYS is a program that provides current
degree-day summaries for thirty-seven sites.
It also generates a three-day forecast for
degree-day accumulation.

PRECIPSUM is a program that provides daily, weekly,
and seasonal precipitation summaries for thirty-
seven sites.

WTHRINPUT is a feature used to obtain detailed environmental data summaries for thirty-seven
sites.

Table 2. A Typical Summary of Sugar Beet Pests as Sent from
a Terminal in a County Agent's Office Showing Current Pests of
this Crop for the Week Ending August 7, 1976

Week Ending:	Aug. 7				
No. of Samples	119				
	0	1-2	3	4-5	
Wireworm	119	0	0	0	
White Grub	119	0	0	0	
Cutworm	119	0	0	0	
Flea Beetle	116	3	0	0	
Springtail	117	2	0	0	
Mines	111	8	0	0	
Leaf Miner Eggs	119	0	0	0	
Var. Cutworm	119	0	0	0	
Armyworm	119	0	0	0	Damage Level
Blister Beetle	110	9	0	0	0 - None Present
Weevil	119	0	0	0	1-2 - Below Economic
Aphids	109	10	0	0	Level
Plant Bugs	82	35	2	0	3 - At Economic
Leafhopper	99	20	0	0	Level
Root Aphid	119	0	0	0	4-5 - Above Economic
Petiole Borer	101	18	0	0	Level
Powdery Mildew	119	0	0	0	
Leaf Spot	119	0	0	0	
Rhizoctonia	62	54	0	3	
Black Root	118	1	0	0	
Number of Samples With Nematode Symptoms			3		
Do You Want to Send a Message?		No			

SUMMARY AND CONCLUSION

The complexity of a typical pest ecosystem, with populations whose dynamic behaviors are strongly coupled, dictates
the need for quantitative models if multifactor control procedures are to be used in a single management program. Only
with the aid of these models can we hope to provide a dynamic
description of interactions of the pests, parasites, and host
crops with the capability of utilizing the many variables such
as age structures, time synchronies, and temperature-dependent
rate functions. These systems models are predictive in that
they provide the basis for computing future estimates of population densities, pest damage, and crop yield from current
measurements. However, they are complex, and their development and use require modern computer technology. If these
models are to be used, computer technology is required for
each of the other components of pest management, environmental
monitoring, biological monitoring, and the information delivery system.

Pest management programs relying on empirical models that
do not account for variations in weather and field conditions,
biological changes, effects of varying age structures, and
time synchronies of the interacting populations, etc., are
inadequate for anything more than single-factor control programs, usually the timing of spray applications.

In discussing the role of computers in pest management we
have presented pest management as an on-line control process.
The need for computer technology with this particular approach
to pest control is great, perhaps as necessary as in any other
area of society. However, in this pioneering activity in
computer applications it is refreshing to note that the overall requirements for the on-line management of crop pests
clearly illustrate the vital role computers will play not only
in pest management, but in environmental management. The
complexity of almost any ecological community is such that
sophisticated, computer-based models will be an essential
ingredient in on-line control of all environmental problems.
Furthermore, because of the stochastic nature of the environment, the models will have to be routinely updated with current data and continuously resynchronized with the actual ecosystem. Vast amounts of data will be collected, analyzed,
stored, and automatically inserted into the models. Clearly
on-line pest management is a prototype of on-line crop management and, larger still, on-line environmental management.

REFERENCES

1. Koenig, H.E., D.L. Haynes, P.D. Fisher. 1976. Management of the environment. pp. 344-81 *in* Computers and Public Administration. S.J. Bernstein, ed. Pergamon, New York.
2. Huffaker, C.B. 1972. The principles, strategies, and tactics of pest population regulation and control in major crop ecosystems. U.S./IBP Project Report, December 2:313.
3. Nix, H.A. (ed.) 1971. Quantifying ecology. Ecol. Soc. Aust. Proc. 6:243.
4. Ruesink, W.G. 1975. Analysis and modeling in pest management. pp. 353-76 *in* Introduction to Pest Management. R.L. Metcalf and W.H. Luckmann, eds. John Wiley, New York.
5. Ruesink, W.G. and M. Kogan. 1976. The quantitative basis of pest management: sampling and measuring. pp. 309-51 *in* Introduction to Pest Management. R.L. Metcalf and W.H. Luckmann, eds. John Wiley, New York.
6. Tummala, R.L., D.L. Haynes, B.A. Croft. (eds.) 1976. Modeling for Pest Management: Concepts, Techniques and Applications. Michigan State University, East Lansing.
7. DeWit, C.T. and J. Goudriaan. 1974. Simulation of Ecological Process. Center for Agricultural Publishing and Documentation. Wageningen, Netherlands.
8. Haynes, D.L., R.K. Brandenberg, P.D. Fisher. 1973. Environmental monitoring network for pest management systems. Environ. Entomol. 2:889-99.
9. Fulton, W.C. and D.L. Haynes. 1976. The influence of population maturity on biological monitoring for pest management. Environ. Entomol. 6:174-80.
10. Croft, B.A., J.L. Howes, S.M. Welch. 1976. A computer-based extension pest management delivery system. Environ. Entomol. 5:20-34.

DISCUSSION

R. RILEY: To what extent are you getting economic input into your modeling?
D.L. HAYNES: We have just completed a cost analysis of the modeling program, but we haven't figured the benefits yet, or incorporated anything into the program.
R. RILEY: John Porter from Agway was relating how their pesticidal representatives are taking computers into the field and discussing economic alternatives with the growers. This technology is in the implementation stage.
D.L. HAYNES: We have found certain training programs to be

very useful, such as "gaming," or ecological simulations of
different situations. For example, we have a game called
"Resistance." These have been very successful.

R. ROVINSKY: Your method is multidisciplinary, to get away
from the limitations of individual disciplines. We have found
that such linked systems have their own special problems; this
is especially true of groups involving extension workers and
researchers.

D.L. HAYNES: Yes, it is true that linkages can be a discipline
in itself. For that reason we find out students are more
qualified than the teachers. Also, concerning interdisciplinary work, I think that most of the talk at universities and
USDA is only lip service. Actually the rewards, including
tenure, promotion, and raises, go to the faculty member who
pays lots of attention to *one* discipline and *one* administrator. We need more external evaluation in the case of interdisciplinary faculty. Also, when "task forces" are created,
we need to make sure they have a limited mission and a limited
lifetime.

R. RILEY: I'd like to add, CSRS has a small grant program to
draw up guidelines for interdisciplinary research in the
transitional area between basic research and application or
implementation. We have received a large number of excellent
proposals, many more than we could fund. This sort of funding
pattern can help bring scientists together.

J. GOOD: Information systems such as weather data should be
widely available. Will such systems be provided by federal,
state, or private sources?

D.L. HAYNES: I think these programs should build on existing
implementation systems. Slow changes can be very successful
in improving the system. We need a hierarchical information
system, forming a whole complex of back-up systems to provide
the farmer with the data he needs.

UNIDENTIFIED: What do you see as the impact of the low-cost
microcomputer?

D.L. HAYNES: I see it as being the heart of this hierarchical
system.

UNIDENTIFIED: Who will maintain this system and who derives
benefits from it?

D.L. HAYNES: I would assume that anyone who is a user is also
a biomonitor; to get useful information out of this system
you have to put a minimum amount of biological information in.

UNIDENTIFIED: Do you see this system as supported by the
state?

D.L. HAYNES: The pest management specialist oriented toward
regional control has no other way of deriving an income. Local
funds could not be used in these situations.

Z. WILLEY: Are the models you have developed adequate for

day-to-day decision making by the farmer, and must each farmer
have data on his own field to use the model?
D.L. HAYNES: Yes, all we need are inputs into the general
model; the farmer inputs the insect levels in his own field.
UNIDENTIFIED: Wouldn't the crop mix and history of his own
fields make a difference?
D.L. HAYNES: Yes, they would make some difference, especially
concerning weed control, where the weed seeds present in the
soil are very important and therefore field history and long-
term planning are necessary for nonherbicidal weed control.
B. DAY: The Russian conservation service told us that they
could predict pest populations with a 95% accuracy one year
in advance. Can you do that well?
D.L. HAYNES: One year's population is highly correlated with
the last year's, so it is easy to predict populations on a
large scale. But it is impossible to predict the area-to-area
variation, which is the useful information for control
purposes.

THE STATUS AND FUTURE OF CHEMICAL WEED CONTROL

Boysie E. Day

Department of Plant Pathology
University of California
Berkeley, California

Since the beginning of agriculture, the nature of crop production has been largely determined by the operations necessary for weed control. Weeds were controlled by tillage and the tillage pattern determined the agronomic cycle. We speak of "cultivated lands," a term describing weed control practices, as synonymous with lands used for crop production. Thus historically, the crop production plan has been integrated into the weed control plan. The primary weed control implements, the hoe and the plow, are the very symbols of agriculture. The traditional crop cycle begins with plowing the soil, and harrowing or dragging to break the clods and form a seed bed. After planting, typically a half dozen or more weedings by the hoe or cultivator were required before harvest. This cycle of weed suppression and crop production has prevailed since ancient times. In the process we have progressed from hand work, to animal power, to the diesel engine. However, in recent years an entirely new dimension, chemical technology, has entered the scene and is in the process of revolutionizing weed control and profoundly altering agronomic systems. This is the use of herbicides for weed control supplemental to or alternative to tillage.

Although the minor uses of herbicides have a long history, the modern era of chemical weed control began about thirty years ago with the marketing of 2,4-D and related phenoxy herbicides. There are presently over 150 chemicals employed as agricultural herbicides (1). In 1976 the sale of herbicides in the United States amounted to $1.4 billion (2). This represented an increase of about 5% over 1975. The trend in recent years has been for the dollar value of herbicides used in the United States to double every 3 to 4 years and the tonnage used to double about every 5 years (3). The present outlook is for continued growth but perhaps at a slower rate than in the past (4). Herbicides represent about two-thirds of the pesticide market, that is about twice as much as the combined total of fungicides, insecticides, nematicides, rodenticides, and miscellaneous products. The use of herbicides is primarily agricultural; however, there is extensive nonagricultural use on urban and industrial lands and on highways and utility rights-of-way.

In view of this spectacular increase in the use of herbicides it is appropriate that we reexamine the alternatives in weed control, review the current impact of this new technology on agronomic practice, look at some of the side effects of past and present technology, and speculate about the future.

THE OBJECTIVES OF WEED CONTROL

I begin with the premise that essentially weed-free monocultures rest on a sound basis as the preferred system of crop production and that the creation of such weed-free conditions is a legitimate objective in farming. We hear arguments that such cultures are ecologically unsound, prone to epidemics of pests and diseases, and devoid of biological balance and stability. Crop monocultures are indeed highly artificial and unstable, and demand strenuous effort to create and maintain. Yet, there is overwhelming evidence that the result is worth the effort. It is well established that on lands of reasonable productivity the cost of intensive weed control is amply repaid. On lands of low potential yield the contrary is true, and the focus of the arts of forestry and range management is to deal with mixed stands of vegetation as best they can. A separate case is the crop polyculture where two or more crops are grown together. The need for weed control in such cultures is as great as for monocrop cultures.

THE ALTERNATIVES

There are currently two basic means by which the farmer can achieve essentially weed-free conditions in cropland: tillage and herbicides. These basic methods are supplemented and modified by a variety of preventive, biological, and ancillary managerial practices, but the basic decision is whether to plow or to spray. With modern machinery and selective chemicals each choice can be highly effective and each can be harmful in its own way. In the United States we are currently at the halfway point. Herbicides serve as a supplement to tillage in some instances but in others have replaced tillage. The trend to substitute herbicides for cultivation is in full swing and the end point is by no means clear. In this regard let us examine the advantages and limitations of mechanical and chemical methods.

Motorized tillage is remarkably efficient in destroying stands of weeds prior to planting, disposing of the crop and weed residues, and keeping the space between the seed rows free of weeds. Skilled operators can cultivate to within an inch or so of the seedling crop, removing all but a narrow band of weeds. The critical limitation on cultivation, however, is that it cannot remove this band of weeds, including the highly competitive ones, immediately adjacent to the crop seedling. This is the critical flaw of mechanized tillage. It lacks ultimate selectivity. With some exceptions weeds in the seed row can be removed only by hand at great cost. Thus, the traditional practice has been to cultivate between the rows and hoe within the seed row one or more times per season.

It was formerly believed that frequent tillage of the soil directly benefited the crop. The present evidence is that the principal and often sole benefit of soil disturbance is the reduction of weed competition, and that the effects are otherwise generally more harmful than beneficial to the crop (5). In the case of poorly drained soils and perhaps other instances there may be direct benefits to the crop from plowing; however, these conditions are not predominant in agriculture (6). Tillage is conducive to soil erosion by water and wind. It has adverse effects on soil structure leading to increased compaction and increased runoff of water. Tillage is wasteful of fuel and has a higher labor requirement than chemical techniques.

Tillage is advantageous in pest and disease control for the reduction and burial of crop residues that can carry infestations over to the subsequent crops. Not the least advantage is the positive emotional impact of cultivation. The control of weeds by cultivation is immediate, obvious, and satisfying. Reinforced by millenia of tradition and extolled in verse and song the newly tilled soil appears cleansed, refreshed, and renewed in productivity. The reverse emotional impact is associated with untilled ground and particularly with the dead and dying vegetation following use of herbicides. The evident high toxicity of herbicides to plants leads to suspicions that they endanger other forms of life and are unclean and noxious.

The principal advantage of herbicides lies in their selectivity. Unlike machine cultivation they can selectively remove weeds from the seed row. Herbicides reduce hand labor, save fuel, and shorten the time lag between crop cycles. When used to replace tillage they can prevent the destructive effects of excessive cultivation. The principal limitation of herbicides is that even the best of them cannot fully discriminate between crops and weeds and fully suppress the array of weeds that can be present. There are also a number of

undesirable environmental side effects, principally on crops
and other nontarget plants. These appear to be of minor importance when normal standards of practice are maintained.
    There is nothing to require that farmers choose tillage
to the exclusion of herbicides or the contrary. The rational
procedure is to use each in its proper place in a coordinated
way. The decision, however, requires a clear understanding of
weed ecology and the nature of the response of weed communities to control measures.

WEED ECOLOGY

    Weed-free monocultures are never achieved in practice.
Any practical system of production is unable to create conditions uniquely suited to a crop species such that no other
plants can survive. On the contrary, when all vegetation is
removed and the crop is planted, an open habitat is created.
This initiates the classical steps of ecological succession
whereby bare areas are first colonized by pioneer species followed by secondary and tertiary invaders until a stable plant
community is reached. Intensive management can arrest the
process of succession at the stage of initial invaders, usually annual species. Thus a particular crop, environment, and
cultural system produces not a monoculture but an agronomic
crop-weed flora that is relatively stable so long as a given
set of practices prevail. Under such conditions, as the weed
flora becomes more firmly entrenched, its competitive effect
on the crop becomes an increasing burden. The buildup is the
product of static or repetitive conditions. Whether or not
the stable conditions are chemical, physical, or biological is
less important than the fact that they are relatively constant. Continued dependence on a particular method of seedbed
preparation, system of cultivation, herbicide, or combination
of herbicides will create a community of weeds suited to those
conditions. This is the central fact of ecology that we must
deal with in the management of crop-weed ecosystems, whether
we are concerned with monocultures, pastures, ranges, forests,
or other lands. Each management program, in suppressing one
weed flora, is an act in stimulation of a new one. This is
particularly true of selective chemicals that fail to control
certain weeds, but also applies to the tillage of fields infested with weeds that resist cultivation.

STRATEGY

    The lesson of weed ecology is that weed management

systems must employ a variety of practices in a coordinated way. Prolonged dependence upon a single chemical or cultural practice or repetitive combination of factors, however suc- cessful initially, will fail over the long term. The basis for all effective systems must be rotation. The classical solution has been crop rotation. This involves simultaneous alteration of a number of cultural conditions. Rotation from row crops to pastures and back to row crops or from crops to fallow are time-honored sequences of recognized effectiveness. The more drastic the rotation the greater the impact on the weed flora. For example, the rotation of rice lands to upland crops and back to rice is a powerful means of suppressing aquatic weeds on the one hand and terrestrial ones on the other. Changes in grazing or fallow procedures, in methods of irrigation or fertility practices, and other components of rotation schemes have their effects on weed floras.

Each new herbicide, advance in tillage method, or other increment of technology adds a new dimension of maneuver in rotation schemes, permitting a greater variety of combinations of conditions affecting weeds. Progress in weed control is made through new discoveries that increase the number, diver- sity, and power of methods available for incorporation into weed control and crop production systems. The development of precision motorized tillage machinery and powerful selective herbicides has contributed to a decline in crop rotation as a weed control technique. The farmer often has the option to rotate herbicides or cultural conditions rather than rotating to a secondary, low-income crop.

CURRENT TRENDS

Initially herbicides were employed in a supportive role to supplement tillage. A full schedule of plowing, harrowing, furrowing, and postemergence cultivation was followed and her- bicides were employed for enhanced effect, particularly to control weeds in the seed row. As experience increased and more effective and diverse herbicides became available, it became attractive to eliminate most or all postemergence cul- tivation. This became possible largely as a result of the de- velopment of broad-spectrum preemergence herbicides such as atrazine and trifluralin. Most major row crops in the devel- oped nations are currently grown under reduced tillage proce- dures involving plowing, seedbed preparation, preemergence herbicide treatment, and in some cases selective postemergence herbicide applications. Postemergence tillage has been essen- tially eliminated in many crops. The widespread adoption of such methods has accounted for the spectacular increase in

herbicide usage over the past 15 years. The success of reduced tillage is dependent upon the availability of a diverse
arsenal of herbicides to meet the rapid changes in flora
generated by the imperfect selectivity of available chemicals.
Availability of a dozen or so selective herbicides with diverse properties might be looked upon as necessary for full
chemical weed control in a crop.

There is a common notion that evolved resistance of weed
species is a major problem with herbicides. This is rarely
the case. The shifts in vegetation are primarily ecological
rather than genetic. There are always other species tolerant
to a chemical that replace the susceptible one, thus reducing
the pressure for evolution of resistance. There are, however,
known cases of evolved resistance and given enough time, perhaps decades or centuries, this could become a significant
factor.

The elimination of tillage after planting could seemingly
represent some end point in the agronomic exploitation of
herbicides. It is a system that supplements and augments
time-tested tillage practices with powerful chemical techniques in mutually supporting ways. All would appear to be in
harmony with the currently recognized need for the coordinated
employment of diverse methods under the rubric of integrated
pest management. However, such is not the case. Doubts have
arisen about the value of plowing the previous crop under and
preparing a new seedbed. Could not these operations be replaced by the use of chemicals? The present evidence is that
in many cases they can be. This opens the prospect that chemical weed control could go all the way and eliminate virtually
all of the remaining uses of tillage.

NONTILLAGE OR CONSERVATION TILLAGE

The nontillage, no-till, zero-tillage or spray-plant-har-
vest system is the culmination of the chemical revolution in
row crop production. The new crop is planted directly into
the stubble or residue of the old crop. There is no plowing
or seedbed preparation. A planter is employed that opens a
slit in the soil, inserts the seed, and presses the slit back
together again. Contrary to popular notions, most crop seeds
do not require a finely prepared seedbed. Weeds are controlled entirely by chemicals prior to and during the crop
cycle and often during fallow between cycles. Such systems
have been successfully applied to corn, cotton, peanuts,
wheat, tobacco, grain sorghum, and certain forage and vegetable crops (6). Corn is the crop most widely grown under
nontillage.

The principal limitation on the method is not adverse soil conditions or a faulty crop environment but the imperfections of available herbicides. Since no single herbicide is able to kill all weeds capable of infesting a crop, a variety of herbicides must be available to meet the changing weed populations that arise. Suitable chemicals are not always available. Since the farmer is entirely dependent on herbicides for weed control, he must carefully fit his treatments to local weed floras and environmental conditions. His arsenal of herbicides may not be sufficient to do the task. This is a deterrent to the expansion of nontillage in row crops, a problem that is gradually being remedied as new herbicides are developed.

Nontillage systems effectively eliminate wind and water erosion. Soil erosion may well be the most serious of all environmental problems. This alone would seem to be sufficient reason to opt for chemical weed control, even at the cost of government subsidy. Erosion is essentially eliminated because living plants or plant residues cover and retain the soil at all times. This increases water infiltration and reduces evaporation, resulting in more water for plant growth. Where water is limiting, crop yields are increased by conservation tillage. Thus, marginal dry lands can be more effectively farmed. Also hill lands not farmed because of erosion can be put into production.

Time-consuming land preparation between crops is eliminated. The time saved makes it possible to grow two crops per year in many temperate regions where only one could formerly be grown. In some cases the new crop can be interplanted in the old one. In this way, for example, soybeans can be planted in the cereal grains before the grain ripens. When the grain is harvested there is an "instant crop" of soybeans with time to mature before the end of the season. Nontillage systems offer greater flexibility in land use and present challenging possibilities in crop production yet to be tested. No doubt many problems will be created or aggravated by the new production techniques. Control of insect pests, plant diseases, and rodents could become more difficult. The management of poorly drained and waterlogged soils could become a problem, and there are other problems, some known now or suspected and others yet to be discovered. We cannot expect to abandon production systems that have proven themselves over the centuries without encountering unsuspected problems. Some of these may be beyond our immediate capacity to cope with and may require temporary retreats to traditional methods. However, the substitution of chemical methods for tillage continues apace and shows no sign of abating.

Lower production costs continue to be the driving force

in the shift to chemical weed control. The initial savings
from eliminating hand tillage were very great. The later reductions in machine tillage have brought important but less
dramatic economies. The economics of the complete elimination
of tillage still appear to be favorable but the margin may be
small. Fuel consumption by machines is reduced by four-
fifths, but at a cost of chemicals that may be increased two-
or three-fold (5). The economics are such that minor factors
could tip the balance one way or another. However, the outlook is clear that chemicals will henceforth be the predominant means of weed control with tillage playing a supplemental
part or perhaps little part at all.

POLYCROP CULTURES

　　We find less need for the weed-free monoculture as we
leave row crops and examine the vegetation management needs
of pastures, ranges, orchards, vineyards, and forests. This
is not to imply a lesser need for an appropriate level of weed
control, but to recognize that something less than a full
monoculture may be the proper objective.
　　In orchards and vineyards plowing is always more or less
destructive to the root system of the crop. Plowing also
tends to create an impervious layer of soil called the plow
sole that is harmful to the trees and vines. On the other
hand where moisture is plentiful, low-growing vegetation in
the orchard is not highly competitive with the crop. In such
situations mowing is often the preferred method, although this
seemingly ideal solution is accompanied by a long array of
disadvantages dependent upon circumstances. Not the least of
these is the difficulty of controlling rampant vegetation immediately adjacent to the trees or vines. This has traditionally required hand work but, because of the high cost, it is
now common practice to use herbicides around the trunks of
trees or on a strip along the vine rows in grapes and berries.
There is an increasing trend toward treating the entire orchard and eliminating plowing or mowing, because it is cheaper
to use one or two soil-residual herbicide treatments annually
rather than several mowings or cultivations.
　　Citrus, avocados, olives, and other subtropical tree
fruits are currently produced almost entirely under chemical
nontillage with the orchard soil maintained completely bare of
all plant growth. Other trees and vines are more or less under chemical weed management. The trend is toward substituting herbicides for both cultivation and mowing.
　　Weed management on pastures and ranges and in forests
that are selectively cut is largely concerned with counter-

acting the destructive consequences of selective harvest.
Livestock eat the forage plants and leave the weeds to prosper undamaged. Selective timber harvest has the same effect.
No better system could be devised to convert mixed vegetation
to weed patches. Various management techniques, usually some
form of conservative utilization, are employed to reduce to
some extent the destructive effects of selective harvest.

The yield of forests and ranges has usually been too low
to support mechanical and chemical means of vegetation management, although fire and biological controls have been of some
value. 2,4-D and related chemicals have the property of
killing or injuring most broadleaved weeds and hardwoods in
grasslands and coniferous forests without significant injury
to preferred species. There is a recent upsurge in the use of
this technique. There is also an increasing variety of new
herbicides for forest use. With increasing population and increasing demands for animal products and forest products we
can expect an expanded effort toward forest and range improvement through the use of herbicides.

In conclusion, it is evident that agriculture is presently involved in a major adjustment from crop and soil management by machinery to chemical management. Fertility control
is already recognized and practiced as a fully chemical art.
It is evident in many situations that weed control can be put
entirely on a chemical basis with the elimination of tillage.
The full array of chemical tools needed to do this in a wide
range of crops is currently in the process of development.
The crop cycle is becoming increasingly a matter of spray,
plant, and harvest.

REFERENCES

1. Weed Science Society of America. 1974. Herbicide Handbook
   of the Weed Science Society of America. 3rd Ed. WSAA Herbicide Handbook Committee, Champaign, Ill.
2. Ernst and Ernst. 1977. Industry Profile Study 1976. Natl.
   Agr. Chem. Assoc. March. Mimeo.
3. ————. 1976. Industry Profile Study 1975. Natl. Agr. Chem.
   Assoc. April. Mimeo.
4. Andrilenas, P.A. and T.R. Eichers. 1977. Evaluation of
   pesticide supplies and demand for 1977. U.S. Dep. Agr.,
   Agr. Econ. Rep. No. 366.
5. Soil Conservation Society of America. 1977. Conservation
   tillage: problems and potentials. Spec. Publ. No. 20.
6. Triplett, G.B. Jr. and D.M. Van Doren, Jr. 1977. Agriculture without tillage. Sci. Am. 236:28-33.

DISCUSSION

E. DECK: Could you say something about controlling brush and
poisonous weeds in rangelands, and weed trees in forests?
B. DAY: We are dealing in these situations with the management
of mixed vegetation. Much of this land is subject to "selective harvest": the rancher is really just a grass farmer who
uses a machine, the cow, which harvests the grass and leaves
the weeds. The same is true in forest lands; the preferred
species are harvested. But in most cases, because of the low
return per acre, it is not economical to treat these lands,
even with the cheapest herbicides. With increasing pressure
on the land, however, chemical treatment will become economically attractive over the next 10 or 15 years.
M. EDEY: It sounds like what you are promoting here is a new
panacea, but several problems with extensive herbicide use
leapt to my mind: (1) long-term health effects of eating
plants that have been routinely treated with massive doses of
herbicides; (2) the possibility that the weeds may become resistant to herbicides; (3) the long-term impact of herbicides
on the soil ecology, perhaps to soil aerators; (4) destruction
of the habitat of natural predators; and (5) an increase in
insect populations because of herbicide use.
B. DAY: First, I don't know who is promoting herbicide use.
It is mostly, I think, the farmers themselves, not the few
weed researchers. Second, concerning the effects of herbicide use on soil organisms, certainly there are not adequate
studies. Ultimately, whether or not herbicides are used will
be decided by which farmers go broke and which ones prosper.
Effects of herbicides on yield are being studied. Herbicides
have other advantages; for example, you can harvest winter
grain in the spring, and a few yards behind the combine plant
soybeans. In other cases, you can plant a bean crop while
the winter grain is still in the field.
T. CLARKE: What about adverse synergism of herbicides with
fertilizers?
B. DAY: Ordinarily, generalizations concerning herbicides cannot be made. Individual herbicides must be looked at. With
atrazine, there are no problems that I know of. Back to the
question on resistance to herbicides, weeds really are a composite pest. The composite "evolves" resistance, so that you
get a new flora in 3 or 4 years. The ecological forces are
much faster and more powerful than the genetic forces, which
is why you rarely get resistance in specific weed species.
M. JACOBSEN: In view of the possible problems with herbicides,
such as the health effects of dioxins in 2,4,5-T, have there
been studies of possible alternatives to herbicides, such as
biological control?

B. DAY: The only alternative in a monoculture is to plow. Can you imagine a biological agent that would eliminate all the weeds in corn? If you can, you have a world-beater! For range management, however, we predominantly use biological control and management techniques.

R. SWEET: What are the possibilities for manipulating crop varieties to make them more competitive with weeds?

B. DAY: Crop varieties that grow quickly and with a certain architecture can shade out the weeds. The modern corn or cotton plant, for example, can compete well after it is a certain age. However, we cannot imagine a shading variety of onions, for instance. Competitive varieties are generally used along with plowing and chemicals.

PEST CONTROL STRATEGIES: URBAN INTEGRATED PEST MANAGEMENT

William Olkowski, Helga Olkowski,
T. Drlik, N. Heidler, M. Minter,
R. Zuparko, L. Laub, L. Orthel

Urban Biological Control Project
Division of Biological Control
University of California, Berkeley

A need exists for the establishment of integrated pest management (IPM) programs in urban areas to reduce pesticide use. Eight years of work has gone into developing such programs in north central California and the San Francisco Bay area with a good deal of success. This paper serves to document some of our findings.

Attention is focused on urban pest problems, because a high proportion of the U.S. population lives in urban areas. In 1970, 73.5% of the population (149 million) was classified as urban by the Bureau of the Census. These people lived in 7,061 communities, each containing 2,500 or more people.

Historically, insect pest control has developed in relation to agricultural, silvicultural, and medical pest problems, while horticultural pests have been given relatively less attention. This general trend is also evident in the development and application of the new technology of integrated control (1-4). The revised edition of *Urban Entomology* (5) considers few horticultural problems. Those that it does consider are dealt with from the relatively narrow but widely accepted viewpoint of chemical control. Only recently, in 1974, did the American Institute of Biological Sciences recognize the field of urban entomology by providing for a symposium on the subject. In 1976 the International Entomological Society also held a session devoted to this area. Very few researchers are available to focus on the problem of urban plant-insect relationships. Those working in this field are scattered and poorly organized (see the Ornamental Plant Entomologists' Newsletter), as is the literature of horticultural entomology. This situation exists despite the fact

that most people live in the urban areas where such vegeta-
tion-pest systems are evident.

URBAN PESTICIDE USE

     One way to assess existing urban pest management activ-
ities is to examine pesticide use patterns; however, little
information is available in this area. Urban and suburban
pesticide users include all sectors of the community: homeowners, public and private institutions, and commercial estab-
lishments. Pesticides are used wherever people maintain
vegetation or provide other food sources on which insects and
other pests can survive. The fact that almost no work has
been done either to catalog the sources of pest problems in
urban areas or to assess pesticide use rates indicates a
serious flaw in overall pesticide monitoring programs, particularly since the potential for human exposure to toxic
materials is so great in urban areas.
     Attempts that have been made to ascertain the degree of
pesticide usage in urban areas include a study commissioned by
the Environmental Protection Agency (EPA) (6) and two independent surveys in California conducted by volunteers[1,2]. The
EPA reported that pesticide production for the nation in 1970
was 1.034 billion pounds, distributed as 47% insecticides, 39%
herbicides, and 14% fungicides. It was estimated that over
750,000 pounds of active ingredients were used in the three
major metropolitan areas studied. This 750,000 pounds was
distributed as 68% insecticides, 17% herbicides, and 16% fungicides. Thus, insecticides appear to be used in proportional-
ly greater amounts in urban areas (Table 1). By extrapolating
from this assessment, an estimated 30 million pounds of total
pesticide is used in urban areas. Roughly 80% (from Table 1)
is used by homeowners. Although the EPA report identifies
lawns as a major target category, the two California surveys
suggest other substantial uses.
     The first survey was conducted in 1971 by a student in
Berkeley, California who sampled 85 homes located in different
areas of the city. Based on the value of the home, it appeared that people with higher incomes had fewer "pests," used
more insecticides, and had different "pests" from people with

[1]*Survey by an unidentified student, under the direction
of John Laing, University of California, Berkeley, Division of
Biological Control.*
[2]*Survey by the Livermore Ecology Center, under the direction of Dale Hattis (formerly graduate student, Stanford University).*

Table 1. Estimated Quantities of Insecticides Used in the Suburbs of Three Major Cities in the United States, 1971, by Category of Use[a].

Use Category	Dallas Texas	Philadelphia Pennsylvania	E. Lansing Michigan	Totals[b]
Homeowner	86%	85%	64%	84%
	216,000	198,500	14,500	429,000
Pest Control Operator	8%	12%	20%	10%
	19,000	27,440	4,650	51,050
Golf Courses	3%	2%	4%	3%
	7,600	4,600	1,000	13,200
City Parks	4%	1%	not available	2%
	9,500	900		10,400
Mosquito Abatement	not available	1%	1%	1%
		3,000	2,500	5,500
	252,100	234,440	22,650	509,190

[a]Source of data: (6). The three areas have a combined population of about 5.5 million and 1.2 million dwellings, distributed as follows:

	Population	Dwellings
Philadelphia	3,866,000	879,413
Dallas	1,327,000	307,775
E. Lansing	272,000	56,658
	5,465,000	1,243,846

[b]The entire study estimated that slightly over 750,000 lb active ingredients were used totally distributed as: 68% insecticides, 17% herbicides, and 16% fungicides. The total pesticide production for the nation in 1970 was 1.034 billion pounds distributed as: 47% insecticides, 39% herbicides, and 14% fungicides.

lower incomes. However, geographic and microclimatic differences could be partly responsible for this variation. Conversely, people with lower incomes had more "pests," but used less insecticides. Older people (over age 40) appeared to use more pesticides regularly, but sample size was too small to make definitive statements. Most of the people sampled fell into the age group 20 to 40, and interestingly, almost all

(79%) said that recent concern with environmental matters had changed their attitude toward pesticide use. More than 25% of the people interviewed used flea collars, 7% used snail pellets, 6% used chlordane in the garden, and about 15% used aerosols for flies and roaches. Thus, although environmental concerns were expressed, pesticide use was still considerable. In general, upper income homeowners either had a gardener to keep things tidy (insects are untidy), hired a commercial pest control company for regular treatments (usually monthly), or applied the pesticides themselves with advice from various sources. Homeowners hiring gardeners generally did not know what specific pesticides were being used on their premises.

Volunteers conducted a second survey in 1972 in Livermore, California, collecting one hundred sixty-one interviews. In response to questioning about the kinds of pest control measures used in and around their homes, 57 people or 31% said they called in professional applicators, 26% reported using Shell No-Pest-Strips®, and 22% used flea collars. When asked to name the particular pest problems associated with their pesticide use, 22 people (14%) identified insects as the problem. An additional 29 people (18%) specifically mentioned one or more of the following targets for pesticide applications: dormant insects—no particular problem named (5 people), earwigs (4), spiders (3), rust fungi (3), bugs (3), moths (2), ants (2), aphids (2), fleas (2), cutworms (1), "sap bugs" (1), weeds (1), carpet beetles (1), termites (1), lawn—no problem named (1), and others such as spider mites, pine bark beetles, "walnut bugs", and elm leaf beetles (4). Eleven people (6.8%) had their homes fumigated (presumably for termites). Two people specified that pesticides were used as preventive treatments and could not name a target pest. Areas of pesticide use and place of purchase are listed in Table 2. Table 3 identifies sources of information for pest management used by these homeowners.

From these studies it can be seen that urban pest management activities are affected by the following—

1. plant species, their locations, and existing management practices;
2. the home, its structure, upkeep, location, and surrounding vegetation;
3. life style of occupants, e.g., food production, waste storage management practices, presence and care of pets, and/or other domestic animals;
4. income, age, and attitude of users toward insects, pesticides, and environmental and human health concerns;
5. geographical location, climate, soils, native vegetation, industrial pollutants, etc.

Table 2. Place of Use and Purchase of Pesticides, Results of 161 Interviews Collected in Livermore, California, 1972

Place of Use	% of Users	Place of Purchase	% of Users
Garden	60	Nursery	42
Lawn	17	Grocery Store	12
House	10	Hardware	10
Other	3	Other drug discount	19
Undetermined	10	Undetermined	17
	100		100

Table 3. Sources of Information on Pesticide Purchase and Pest Management, Results of 161 Interviews Collected in Livermore, California, 1972

Source	%
Nursery	37
Friend, Neighbor, Relative	10
Other Store	10
Hardware	8
Grocery Store	7
T.V.	6
Newspaper, Magazines, Garden Book	3
Other	3
Undetermined	16
	100

Our experience in developing urban IPM programs has emphasized these conclusions. The problem of pesticide use is multi-faceted, and it has become clear that to deal with it effectively one must use a holistic or systems approach taking into consideration political, sociological, and biological factors.

STREET TREE PEST MANAGEMENT AS AN INITIAL APPROACH

Given the complexity of the urban area with its multitude of people and their diverse pest control practices, widespread fear of insects (7), and pesticide-oriented information sources, where can one start in order to bring the IPM approach to urban populations? Urban street trees, which were selected initially for their high visibility, are responsible

for substantial budgets devoted to pest control, and are
frequently the concern of vocal, politically active citizen
constituencies interested in bringing "new" ideas into the
community. The microgeographical location of street trees,
in the interface between private and public domains, gives
these trees a critical position within the city vegetation
complex. Street trees also provide very practical advantages
to people by reducing the need for air conditioners through
shade and added humidity as well as by improving the liv-
ability of crowded urban areas through aesthetic enhancement.

By setting up IPM programs for city governments rather
than individuals, models are created that can offer viable
alternatives to the homeowner if properly represented to the
public through existing mass media. This can be an energy
cost-effective way to disseminate IPM concepts and technology,
as the individual could not afford the cost of initial program establishment.

When work was started in 1969, the usual approach to pest
control research was used: focusing on a particular pest
problem, investigating its natural history, and conducting
field and laboratory studies. However, soon after the start
of the project, it became obvious that the traditional approach was inadequate and that a systems view of the problem
was needed that included the human community within which the
pest problem had arisen.

The project had begun with a focus on biological control
of the linden aphid, *Eucallipterus tiliae* L. in Berkeley,
California (8). After the first year, city personnel and
citizens who were aware that this initial effort reduced insecticide use on this insect began pressuring for further
study of other pest problems. It became apparent that some
of the problems were related to horticultural maintenance
practices and some were a function of the pest control treatments. Indeed, the primary problem was the pest management
delivery system and the lack of an ecosystem perspective.
Only occasionally was the problem due to lack of information
about the specific pest (with the exception of information on
biological control agents). Results of this process of discovery have been reported elsewhere (9). It is important to
record here the differences between the pest management approach originally developed for agricultural crops and the
pest management approach we have been developing for urban
street trees.

DISTINCTIVE CHARACTERISTICS OF URBAN IPM

Urban IPM (especially but not exclusively street tree IPM) involves more pest problems than are usually encountered in a one-crop agricultural system. In the San Francisco Bay area, there are more than 100 species of street trees commonly planted. Each of the plant species has its own complex of insect pests. Table 4 is a list of the street tree pest problems now being studied in our project. It shows the distribution of one season's requests for pest management advice, received from the communities supporting this work.

The proximity of people to the pest-vegetation complex is another difference between urban and agricultural or silvicultural pest management; therefore, extreme caution should be exercised in the use of toxic materials. Wherever possible alternative strategies such as physical, cultural, and biological controls should be employed.

Characteristically, urban vegetation systems are maintained primarily for aesthetic purposes. Most of the pest problems encountered by this project are not economically damaging to the vegetation; thus economic pressures to treat insect populations are frequently not a factor in the decision to use insecticides. Exceptions do exist; for example, Dutch elm disease has devastated elms across the country, costing municipalities thousands of dollars.

THE INJURY LEVEL CONCEPT

The existence of a hypothetical aesthetic injury level has been assumed (8-10) because it helps to explain treatment of insect populations when no economic damage is apparent. Application of this concept to field situations can greatly reduce pesticide treatments.

An example is found in the work on the California oak-moth, *Phryganidia californica* Packard (11), an occasional defoliator of the California live oak, *Quercus agrifolia*. Periodic defoliation of this native tree by this native insect is not permanently damaging. A sampling program was developed for this insect, and after studying several populations, an aesthetic injury level was set at 10 larvae per 25 shoots. Above this level excessive defoliation occurred. Although no extensive population data were correlated with damage levels, the aesthetic injury level was useful in determining when to treat in order to prevent the temporary loss of leaves regarded as unsightly by many people.

In 1976, 40 European holly oaks, *Q. ilex*, were monitored in Berkeley, California by our urban IPM group using the

Table 4. A List of the Pest Problems Under Study by the Urban Biological Control Project, University of California, Berkeley, California, 1977 Showing Monthly Distribution of Requests for Pest Management Advice from Cities Participating in the Program During 1976 (Note, this list does not include indoor pest problems studied by the UIPM project in the Palo Alto School District, Palo Alto, California[a].)

Pest Species	Importance Rank	Importance Total	%	Jan	Feb	Mar	Apr	May	Jun	Jul	Aug	Sep	Oct	Nov	Dec
Phryganidia californica California oakmoth	1	575	46	-	1	24	76	386	40	12	11	7	15	3	-
Hyphantria cunea Fall webworm	2	123	9.8	-	-	-	-	-	4	72	38	7	2	-	-
Pyrrhalta luteola Elm leaf beetle	3	98	7.8	-	-	-	1	3	15	25	20	22	10	2	-
Drepanaphis acerifolii Silver maple aphid	4	72	5.8	-	-	-	8	7	1	2	8	42	4	-	-
Illinoia liriodendri Tulip tree aphid	5	65	5.2	-	-	-	2	24	19	6	11	1	2	-	-
Archips argyrospilus Fruit tree leaf roller and *Lithophane attenata* Green fruit worm	6	57	4.6	-	-	-	42	15	-	-	-	-	-	-	-
Prociphilus fraxinfolii Modesto ash aphid	7	38	3.0	-	-	-	5	8	16	6	1	1	1	-	-
Euceraphis punctipennis and *Callipterinella calliptera* Birch aphids	8	23	1.8	-	-	1	5	10	3	2	-	2	-	-	-
Schizura concinna Red-hump caterpillar	8	23	1.8	-	-	-	-	1	1	2	7	12	-	-	-
Myzocallis sp., *Tuberculoides* sp. and *Tuberculatus* sp. Oak aphids	10	18	1.4	-	-	1	5	7	3	-	2	-	-	-	-
Orgyia vetusta White faced tussock moth	11	17	1.4	-	-	-	4	12	1	-	-	-	-	-	-

Hyalopterus pruni Plum aphid	12	14	1.1	-	-	1	2	9	1	1	-	-	-	-	-
Gossyparia spuria European elm scale	13	12	1.0	-	-	1	2	6	1	2	-	-	-	-	-
Malacosoma sp. Tent caterpillar	14	11	.9	-	-	1	5	2	2	1	-	-	-	-	-
Periphyllus lyropictus Norway maple aphid	15	10	.8	-	-	-	-	4	1	1	1	3	-	-	-
Eucallipterus tiliae Linden aphid	15	10	.8	-	-	-	2	-	2	3	1	1	-	1	-
Tropidosteptes illitus Ash bug	15	10	.8	-	-	-	7	1	-	2	-	-	-	-	-
Saissetia oleae Black scale	18	8	.6	-	-	-	2	5	1	-	-	-	-	-	-
Leptocorris rubrolineatus Western box elder bug	19	7	.6	-	-	-	-	2	1	-	1	2	1	-	-
Scolytus rugulosis Shot hole borer	20	5	.4	-	-	-	1	-	1	1	1	1	-	-	-
Tinocallis plantani Elm aphid	20	5	.4	-	-	-	-	1	1	2	1	-	-	-	-
? Hackberry aphid	22	4	.3	-	-	-	-	-	1	2	1	-	-	-	-
Trialeurodes sp. Whiteflies	22	4	.3	-	-	1	-	-	1	-	1	-	-	1	-
Calico sp. Sweetgum scale	24	3	.2	-	-	-	1	-	1	-	-	-	1	-	-
Hordnia circellata Blue-Green sharpshooter	24	3	.2	-	-	-	-	-	2	-	1	-	-	-	-
Miscellaneous aphids	-	21	1.7	-	-	2	2	3	2	7	1	1	2	1	-
Miscellaneous lepidoptera	-	5	.4	-	-	-	-	1	-	2	2	-	-	-	-
Miscellaneous scales	-	2	.2	-	-	-	-	2	-	-	-	-	-	-	-
Miscellaneous	-	8	.6	-	-	-	2	1	-	-	2	3	-	-	-
Total number of requests		1251	100%	ϕ	1	32	174	510	121	151	111	105	38	8	ϕ
Percentage of total			100%	ϕ	.1	2.5	13.9	40.7	9.7	12.1	8.9	8.4	3.1	.6	0
Cumulative percentage			100%	ϕ	.1	2.6	16.5	57.2	66.9	79.0	87.9	96.3	99.4	100	100

[a] See Table 5 for additional information on the cities participating in this project and their respective sizes.

monitoring system developed by Pinnock and Milstead (11).
The aesthetic injury level was reduced to 8 larvae per 25
shoots because the holly oak was determined to be more
vulnerable to oakmoth injury than the native *Q. agrifolia*.
These trees were young, small, and situated such that
defoliation was extremely noticeable. By monitoring and
application of the aesthetic injury level discussed above,
only 30% (12 out of 40) of the trees required treatment[3] to
prevent possible defoliation (fig. 1). Ordinarily all 40
trees would have been treated when the caterpillars were
noticed.

In the spring of 1977, this injury level was applied on
a larger scale when 5,680 oaks (mostly *Q. ilex*) were monitored in San Jose. Because of the extent of the system and
limited time for decision making, the oakmoth larval populations were classified as high, medium, or low. High and
medium trees were those that exceeded the injury level and
were consequently sprayed with *B. thuringiensis*. A total of
2,300 or 55% of the trees were treated. Thus, through application of the aesthetic injury level concept, 45% of the
trees were eliminated from the treatments, conserving material and money.

The blue spruce aphid, *Elatobium abietinum* Walker, provides an example of a pest that can cause economic damage.
Successive defoliations of its host plant *Picea* sp. are
believed to kill the tree (12). This aphid occurs on blue
spruce plantings in a highly visible median strip in the city
of Palo Alto. These trees were regularly treated with a
series of different compounds over the years. Just prior to
beginning IPM work, pyrethroids were used on all 30 trees.
In 1976 detailed studies were initiated that documented the
population sizes on all trees. Aphid populations were monitored biweekly by sampling one branch from the north, south,
east, and west sides of the tree, using a beating cloth. The
lowest aphid counts that correlated with defoliation provided
a tentative working injury level of a mean of 34 aphids per
quadrant per tree. During the first study season only 17%
(5 out of 30) of the trees studied showed sample means above
the injury level and were treated (fig. 2). None of the
study trees were defoliated. We plan to continue this study
in order to verify the proposed injury level and to discover

[3]*The material used was* Bacillus thuringiensis, *a bacterial insecticide, which is particularly well suited to urban
IPM by virtue of its order-specific (Lepidopteran larvae)
activity and its nontoxicity to mammals.*

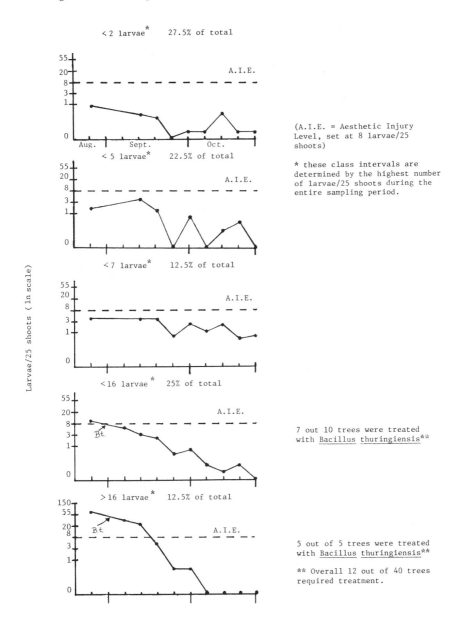

Fig. 1. A display of various classes of abundance of
California oakworm (*Phryganidia californica*) larvae on holly
oaks (*Quercus ilex*) in Berkeley, California, 1976.

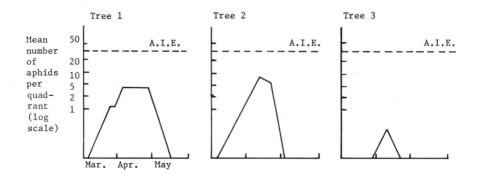

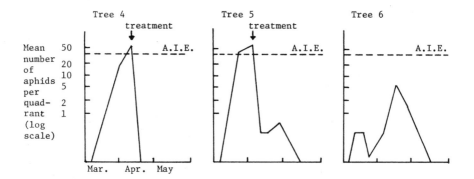

Fig. 2. *Blue spruce aphid populations on 6 trees in Palo Alto, California. Each tree was monitored from March 4 to May 27. Populations on trees 4 and 5 exceeded the aesthetic injury level (A.I.E.) and were treated with Pyrenone®.*

whether individual trees repeatedly suffer excessive aphid populations.

SPOT TREATMENTS

It has been observed that there is often wide variation in pest numbers on adjacent trees of the same species or in the general level of severity of a particular pest in different stands of the same tree species in a given city. We observed this phenomenon in several tree-herbivore interactions (13), but the controlling mechanisms are as yet unknown. The fact that this phenomenon has been observed with a plant pathogen, *Gloeosporium aridum* Ell. and Holw. and with the ash bug, *Tropidosteptes illitus* (Van Duzee) on Modesto ash trees, *Fraxinum velutina* 'Modesto' (a clone), indicates that variation in the environment and not differences in genetic resistance in the host tree may account for variation in the severity of these pest problems. Even without knowing the mechanism of this phenomenon, confining treatments to those trees showing high pest populations year after year can reduce pesticide use.

THE NEED FOR DELIVERY SYSTEM RESEARCH

It is a mistake to assume that delivery system strategies commonly employed in agricultural IPM projects can be transposed to urban areas. The problems of social, political, and biological interactions and complexities in urban areas have already been discussed. These factors directly affect pest management decision making. Furthermore, in an urban IPM program citizens alert pest management advisors of potential pest problems while entomologically untrained municipal personnel are relied upon to carry out necessary management strategies (fig. 3 describes this process in detail). Our project has begun to develop management devices specifically tailored for the urban situation (fig. 4), but more work is needed in this area. Each type of institution has its own idiosyncracies, and as we have learned in our project, a · school district requires quite a different approach than a city public works department. An additional benefit of developing viable urban IPM delivery systems is the potential for making the technology available to the private pest control industry, thus expanding the role of IPM in cities.

INCLUDING BIOLOGICAL CONTROL RESEARCH IN IPM STUDIES

Without the capability to import parasites and predators, IPM programs, particularly those dealing with nonnative pest species, tend to focus on servicing problems rather than implementing permanent solutions. If IPM programs are organized under one institutional framework and biological control importation programs under another, as has been the case in California agriculture, each organizational approach develops independently. The result is that the field-oriented IPM people may forget about the possibilities and importance of nurturing importation projects, while the biological control experts become somewhat isolated from the field work and neglect research into the development of delivery systems.

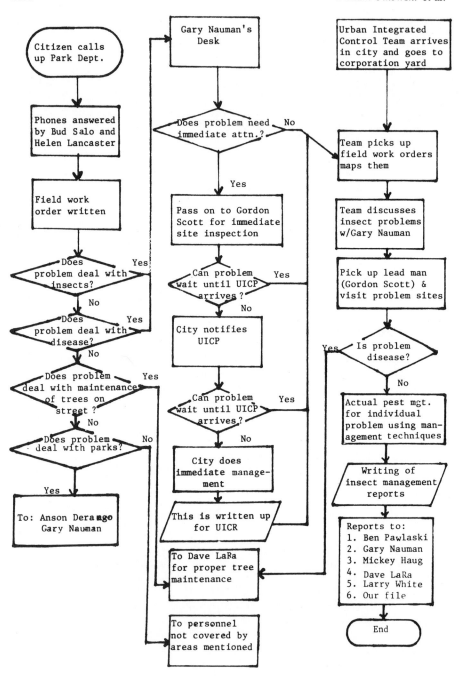

UICP = Urban Integrated Control Project Team

Fig. 3.   Citizen complaint procedure—Palo Alto 1976.

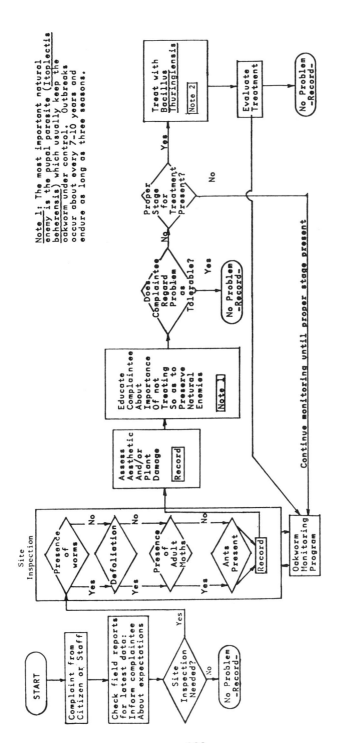

Fig. 4. The California oakworm complaint management process (*Phryganidia californica*).

229

The resulting situation creates unnecessary lags in support for the initiation of natural enemy studies and introduction programs and, obviously in some cases, in pest management strategies. Biological control research is particularly applicable in urban areas, where permissible levels of herbivorous insect abundance may be much higher than in agriculture, thus providing an ideal setting for the use of parasites and predators to suppress pest population numbers.

EDUCATION

Education is an important component of an urban IPM program. In dealing with insect-related service requests from citizens, the concept of injury level, the role and importance of beneficial insects, and the problems associated with the exclusive reliance on synthetic pesticides are usually presented informally. Reprints of articles of interest or one-page handouts developed by our staff that deal with specific pest problems are frequently given to citizens. In Palo Alto, California, one-page information sheets discussing some aspect of IPM (e.g., soap and water washing for control of aphids) are periodically included with the monthly bill from a city-owned utility company. Increased awareness of IPM concepts by citizens can encourage rational approaches in dealing with household pests and insects feeding on privately owned ornamental and food plants. Emphasis should also be placed on fostering in city horticulturalists an understanding of the principles of IPM and knowledge of the biological information pertinent to the management of each pest, in order to improve the ability of city personnel to carry out the work of developing and implementing an IPM program.

CONCLUSIONS

Integrated pest management programs in urban areas can significantly reduce pesticide use in these environments (see Table 5). This can reduce operating costs to city governments and provide a healthier urban environment, while conserving the effectiveness of pesticides and reducing the development of pest resistance to available chemicals.

To be effective, an IPM program must provide the greatest possible range of alternative pest control strategies. Biological control research must be integrated with delivery system studies into a comprehensive program, acknowledging the socio-political environment of urban areas. Aesthetic and economic injury levels must be determined for horticultural

Table 5. Pesticide Reduction Summary of the IPM Shade Tree System on the Urban Biological Control Project, University of California, Berkeley, at Completion of 1976 Season

| City | No. of Years Under IPM Program | City Population Census[a] | Total Tree Pop. Under Mgmt. | Total Insecticide Treatments in Year Prior to Introduction of IPM Program[b] (In number of tree treatments) | 1976 IPM INSECTICIDE TREATMENTS (In Number of Tree Treatments) | | 1976 Total IPM Insecticide Treatments | % of Treatment Reduction Due to Implementation of IPM Program |
					1976 Microbial Treatments (*Bacillus thuringiensis*)	1976 Chemical Treatments (Diazinon, Malathion, Sevin, Pyrenone, & MetaSystox-R)		
Berkeley	6	107,500	35,000	11,500	32	3	35	99.7%
San Jose	3	557,700	250,000	42,000	4,307	3	4,310	89.7%
Palo Alto	2	54,900	80,000	1,600	369	6	375	76.6%
Modesto	1	85,000	85,000	8,000	8	350[c]	358	95.5%
Davis	1	32,800	12,000	10,000	0	12[d]	12	99.8%
Total:		837,900	462,000	73,100	4,716	374	5,090	93.0%

[a] Census, January 1976.
[b] All treatments were chemical (no microbial agents were used).
[c] 4,250 trees were treated with MetaSystox-R (injection) prior to the initiation of IPM program in March, 1976.
[d] 805 trees were treated with various chemicals prior to the initiation of IPM program in April, 1976.

and other urban pests to provide reasonable treatment guidelines for use in monitoring programs. In addition, it must be
recognized that citizen education is a vital component of any
urban IPM program that facilitates the job of a pest management advisor in furthering the recognition of IPM as a viable
alternative to strictly chemical pest control.

REFERENCES

1. Huffaker, C.B. 1971. Biological Control. Plenum, New York.
2. Huffaker, C.B. and P.S. Messenger. (eds.) 1976. Theory and
   Practice of Biological Control. Academic Press, San
   Francisco.
3. Metcalf, R.L. and W.H. Luckmann. 1975. Introduction to
   Insect Pest Management. John Wiley & Sons, New York.
4. van den Bosch, R. and P.S. Messenger. 1973. Biological
   Control. Intext Educational Publishers, New York.
5. Ebeling, W. 1976. Urban Entomology. Agricultural Sciences
   Publications, University of California, Berkeley.
6. von Rümker, R., R.M. Matter, D.P. Clement, F.K. Erickson.
   1972. The use of pesticides in suburban homes and gardens
   and their importance on the aquatic environment. Pesticide
   Study Series 2. Office of Water Programs, Applied Technology Division, Environmental Protection Agency, Washington,
   D.C.
7. Olkowski, W. and H. Olkowski. 1976. Entomophobia in the
   urban ecosystem, some observations and suggestions. Bull.
   Entomol. Soc. Am. 22:313-17.
8. Olkowski, W., H. Olkowski, R. van den Bosch, R. Hom. 1976.
   Ecosystem management: a framework for urban pest control.
   BioScience 26:384-89.
9. Olkowski, W. 1973. A model ecosystem management program.
   Proc. Tall Timbers Conf. Ecol. Anim. Control Habitat
   Manage. 5:103-17.
10. Olkowski, W. and H. Olkowski. 1974. Observations on an
    Integrated Control Program for Urban Ornamental Plants.
    Proc. 1974 California Golf Course Superintendents Insti-
    tute.
11. Pinnock, D. and J. Milstead. 1971. Biological control of
    California oakmoth with *Bacillus thuringiensis*. Calif.
    Agr. 25:3-5.
12. Ohnesorge, B. 1959. Die Massenvermehrung der Kitkalaus in
    Nordwestdeutschland. (The Outbreak of the Sitka-spruce
    aphid in northwestern Germany.) Forstarchiv 30(4-5):73-78.
    [Abstract in Rev. Appl. Entomol. 49(1961):281.]
13. Olkowski, W., H. Olkowski, A. Kaplan, R. van den Bosch.
    1977. The potential for biological control in urban areas:

shade tree insect pests *in* Perspectives in Urban Entomology. K. Koehler, and G. Frankie, eds. Academic Press, New York. (In press).

DISCUSSION

J. ANDREWS: I gather you feel that there are good prospects for reeducating the public concerning cosmetic damage to plants. Does the same hold for food?
H. OLKOWSKI: Yes, definitely. We found in consumer cooperatives in Berkeley that with proper education a large number of people will accept considerable cosmetic damage in their foods.
W. OLKOWSKI: In all cases, however, the people in the field must want to reduce pesticide use, not just the public.
H. OLKOWSKI: Once we've gone into an area, assessed the problem, and éducated people, we usually find that all we have left are a few key pests that require alternative management research. Large numbers of problems drop out within the first few years of the program, because they really are not serious.
A. ASPELIN: Concerning the data gap, EPA's Office of Pesticides is now finishing a study on the use of pesticides in households.
R. RILEY: What do you feel are future needs for data?
A. ASPELIN: The nonhousehold part needs to be covered, including institutional and governmental use of pesticides. No significant efforts are now underway to gather this data.
W. OLKOWSKI: We have been able to get good data on use in city parks by developing an intimate relationship with park personnel. When we ask them how much and what kinds of pesticides they are using, they don't know, or they won't say.
H. OLKOWSKI: But when we get close to them and go in and look at the purchasing records, we discover the discrepancy between what they said they were doing and what they are doing.
UNIDENTIFIED: I have a background in IPM, but I've recently moved to Washington, D.C., and I don't want to live with cockroaches. I don't seem to have alternatives to insecticide use.
H. OLKOWSKI: There are ways of managing roaches to acceptable levels and reducing pesticide use. If you use water and habitat control, paint, caulking, and sanitation, and then use chemicals (we use baited traps), you actually can control cockroaches in your own apartment, regardless of the apartment next door. You have to use an integrated control program for cockroaches, and if you do, it is very effective.
B. KAHN: Can you use the Extension Service in your programs?

W. OLKOWSKI: Very little of the state extension literature on
shade tree pest management mentions IPM, injury levels, or
biological control. They are still recommending aldrin,
dieldrin, chlordane, etc. This has got to change.

Part IV
Obstacles and Incentives

CURRENT STATUS, URGENT NEEDS, AND FUTURE PROSPECTS OF INTEGRATED PEST MANAGEMENT

Carl B. Huffaker

Division of Biological Control
University of California
Berkeley, California

Christine A. Shoemaker

Department of Environmental Engineering
Cornell University
Ithaca, New York

Andrew P. Gutierrez

Division of Biological Control
University of California
Berkeley, California

INTRODUCTION

It is apparent from the papers given in this Symposium that what we mean by "pest control strategies" and "integrated pest management" (IPM) differs considerably among the speakers, and would differ more markedly if the meanings attached to these terms by all those dealing with them were included. We use the term "strategy" to refer to basic philosophy and approach and the term "tactic" to refer to specific methods or supporting techniques required to carry out a basic strategy. The two terms are, of course, interrelated in that a strategy utilizes one or more tactics, but they are not synonymous. There are perhaps three basically different strategies: (1) *prevention* or *eradication,* which does not accept existence of the pest, and aims all tactics at its prevention or destruction; (2) *containment,* which accepts existence

237

of noneconomic densities of the pest and aims its tactics simply to prevention of economic loss; and finally (3) the strategy of *doing nothing,* accepting any pest densities that might develop, relying entirely on natural controls. The latter, as a general strategy for management of all pests, clearly is unacceptable, although large acreages of some crops do not need treatment, and utilization of natural controls is itself a key tactic of *containment* strategy. The strategy of IPM is basically that of containment (1,2). The strategy of prevention, or even eradication, is wise in some situations, the latter mainly for recent arrivals that have not become widely established. The preventive strategy seems more appropriate to plant disease control than to insect control.

Professor Ray Smith has given you a synopsis of the historical development of concepts of pest control in general and particularly of integrated control, a term which has evolved in the United States to IPM. We will not rehash this development or urge on you still again the basic reasons why a new look at our pest control technology is being demanded by society and required for profitable crop production. We assume that enough has been said on this subject to be amply persuasive. We will deal here with the basic requirements for making progress toward developing the necessary technology, how far along we are in doing so, what is still to be done, the problems faced, the greatest needs still to be met, and the prospects for meeting these needs and putting a modern IPM technology into practice on a national scale.

The basic requirement for making progress in developing sophisticated, modern systems of crop pest control is to marshall together a well-funded, cooperative, closely organized, and enthusiastic team of competent researchers and education specialists, representing the major disciplines required for such a program. It has become, we believe, painfully obvious even to our administrators who have inherited the old system of filtering down of their funds and support for pest control research that a fundamental change is needed if we are to overcome the rising dilemma posed in meeting our food requirements, protecting the economic viability of farming, reducing our fossil energy needs, and preserving a healthy environment. It is no secret that a scientist commonly works in relative isolation at his or her own specific task without paying much attention to the consequences of the research on that of another discipline, even a discipline involved with the same crop. Scientists are notorious individualists—we like to do our own thing. This individualism is often fine and creative in terms of development of a specific tactic or solution for control of a single pest species, but the recommendations resulting may cause other pest problems of which the individual scientist is unaware or unable to appraise.

Our main systems of agricultural research support have tended to encourage and entrench this uncoordinated, individualist, piecemeal approach. Government at all levels has not fully faced the need to break away from a system that presents no way to support major nationwide team approaches to crop production. Support, in general, has been through the USDA's Agricultural Research Service (ARS), its Cooperative State Research Service (CSRS), and the various individual states. The ARS program has not brought in the many competent scientists working in universities. The CSRS effort to develop the team approach has met with considerable frustration and lack of vigorous accomplishments because of the established pattern by which funds go to the states, and are then channeled through the administrative structure of each experiment station to the various divisions, departments, and sections to the individual scientists for support, largely of their individual efforts. The Regional Project system of CSRS has been commendable as an attempt to correct this but it has not been effective enough.

The International Biological Program (IBP) and the National Science Foundation, with the Environmental Protection Agency, in 1971-72 stimulated the development of and supported a major block-funded multidisciplinary and centrally managed team effort designed to learn if such a program would be more productive than our old way of doing things. You have heard in this Symposium accounts of the progress of some of the work in several of the program's subprojects: for alfalfa (Dr. Armbrust), for deciduous fruits (Dr. Croft), for soybean (Dr. Newsom), and partly in this program and related programs, the utilization of systems analysis and computer technology in IPM (Dr. Haynes). In addition to these three crop areas, the IBP-initiated program also included subprojects on citrus, cotton, and pines (as affected by bark beetles).

We will not duplicate these already detailed discussions, but will summarize certain developments in the cotton, citrus, and pine bark beetles subprojects, which have made major advances in some respects different from those detailed by previous speakers.

OBJECTIVES OF THE IPM PROGRAM AND ITS APPROACH

The objectives of the NSF/EPA supported IPM program have been as follows.

Generally, the objective is the development of improved, ecologically oriented pest management systems that optimize, on a long term basis, costs and benefits of crop protection. Thus, the project is striving to develop systems of pest control that will return greater *profits* to the producer. At the same time, these systems, by reducing use of expensive and counterproductive disturbing pesticides, would produce other large benefits to society.

Inherent to this general objective is recognition that most crop pests are highly adaptive, highly reproductive organisms and that in nature they are not likely to be eradicated; moreover, they do not cause catastrophic damage to their host plants in natural systems. Thus, we seek to maximize natural control forces: weather, host resistance, and natural enemies.

Although the weather cannot be manipulated directly, we can intensify its harmful effects on the pests and ameliorate those on the natural enemies. The other two factors of nature's great triumvirate, plant resistance and natural enemies, have been taken as the cornerstone of the effort. In addition, efforts have been intensified to find better ways of using chemical pesticides primarily by using nonselective ones in ecologically selective ways. The chemical industry offers little hope of obtaining physiologically selective pesticides.

We note also that the goal of attaining *maximum* yields may no longer be practical for many field crops due to pesticide resistance, excessive costs, and government bans on some pesticides. We may need to seek *optimum yields for best returns*. What is needed for the family farmer may be different from that needed for corporation farming. We may need to put more emphasis on developing systems enabling the family farmer to compete more effectively with the corporation operation.

The specific objectives relating to the tactics that may be employed in the overall strategy are (3) —

1. to develop an explicit understanding of the biological, ecological, and economic processes in crop growth, the population dynamics of the pests and their natural enemies, and the interactions among these processes;
2. to develop alternatives to suppress major pests, especially cultural, biological, and host resistance tactics, which are ecologically compatible and can be expected to reduce the use of broad spectrum biocides and lessen the adverse effects of their use;
3. to develop better methods of collecting, handling, and interpreting relevant biological, meteorological, crop production, and economic data;
4. to utilize systems analysis in the general sense and specific modeling as a central unifying and research-guiding tool in the pursuit of the main goal and its subsidiary goals; and
5. to build models of the crop production and pest

systems, integrate these with economic analysis (impact, etc.), and conduct pilot tests on the use of the combined management models for each crop system.

Huffaker and Smith (3) also noted that the basic procedure has been as follows, although not explicitly so in each case (modified after F.R. Lawson and associates):

1. Separate the *real* pests from those *induced* by insecticides in the different regions involved.
2. Establish realistic *economic injury levels* for the real pests, with appropriate attention to the hidden costs of controls.
3. Separate the *real* pests into those causing intolerable losses, i.e., *key* pests, and those causing only light or sporadic damage controllable by occasional or limited use of pesticides.
4. Identify the key factors (e.g., a key resistant variety, natural enemy, or cultural measure) controlling populations of the *key* pests and measure their effects.
5. Design and test control systems based upon these guidelines in each of the areas where the key pests and/or factors are different.
6. Modify control systems according to time and area conditions and new inputs as the program develops.

THE SYSTEMS APPROACH AND MODELING

At the outset, this project accepted the dictum, "consider the ecosystem" (2); each crop has been explored as an agroecosystem. The complexity of growth processes and impending factors, which involve complexes of pests, natural enemies, and interactions among these factors, and the impact of these relationships, make the decision-making process difficult. Prior to this program there had been little effort to tie pest control in its various phases to crop growth and economic gain or loss in any explicit way. The analysis of all the factors and processes in the culture of a crop, control of the pests, and determination and optimization of cost/benefit relationships, is often too complex for intuitive solutions. What was needed was a new technology that would utilize the power of computers and systems analysis in the various forms that have been used in engineering, industry, and commerce. This technology, as you have heard, soon became the focal point of the effort, determining research priorities and guiding much of the research, assessing results of various processes, practices, and interactions (biological,

physical, cultural, economic, social, etc.) and coupling them
to arrive at appropriate control recommendations. The crop
itself was taken as the central feature, and models of plant
growth, phenology, and yield functions have been developed.
Analyses of insect damage (and to a limited extent, damage by
plant pathogens) have thus been coupled to analyses of processes of plant growth and commodity production.

In our efforts to look at the whole system collectively,
we do center on certain key potential tactics. These are
economic thresholds, cultural means, natural enemies, host-
plant resistance, and selective pesticide use.

Moreover, even though modeling has been taken as a focal
point in the development of our programs, this by no means
implies that basic biological studies on the organisms themselves are neglected. Extensive, detailed biological studies
have been conducted. Use of modeling, on the contrary,
forces us to plan such studies more carefully so that the results can be used to define relationships more precisely and
quantitatively to help us make predictions about pest densities and yield for a variety of weather, soil, or other
conditions. Models help us to choose appropriate pest management tactics for given situations. Lastly, they enable us
to examine a much broader range of combinations of variables
than those that can be examined explicitly by field and
laboratory experimentation.

Because of the large effort required to obtain such information for each combination of variables and for each
strategy or tactic, only a few alternatives can be examined
if the effort is restricted to experimental examination.

Modeling is thus a form of extrapolation. When we consider the full range of alternatives that potentially could
be used in pest control, the number of different strategies
and tactics can be very large. For example, consider a situation in which the decision is whether or not to apply an
insecticide each week for 10 weeks. There are then $2^{10} =$
1024 different insecticide "schedules" possible. It would
be impossible to test each of these possibilities in the
field.

The number of alternatives is even larger when programs
other than fixed-schedule insecticide spray programs are considered. In an IPM program, biological and cultural methods
of control must also be considered. In addition, the decision of when to implement a control method is more likely to
be correct when based on a current assessment of the status
of the ecosystem, including the weather, the vigor and maturity of the plant, the kinds and abundance of natural enemies, and the age and density distributions of the pest
population and its natural enemies.

It is largely to overcome this difficulty that models have been developed to describe crop ecosystems. It is hoped that by varying the inputs to such models, the effects of a large number of pest control options, under a broad range of ecosystem conditions, can be simulated. Modeling is a way of extrapolating from the specific to the more general. The potentials of a much larger number of control alternatives can thus be estimated than would be possible using purely empirical or experimental methods.

Simulation models for use in pest management are based on detailed descriptions of the daily interactions among populations of organisms and their abiotic environments. The structure of the model is based on a qualitative understanding of the dynamics of the ecosystem: Which factors affect growth? What does each category of each species eat? How do pest control measures affect the pest population or the natural enemies? On the basis of this type of information the model's structure is developed. Qualitative knowledge of the way the populations interact influences our choice of which aspects of the ecosystem should be described in detail and which can be aggregated.

The structure of such a model is illustrated in figure 1. In order to predict the effects of weather and pest control measures on yields, it is usually necessary to examine the dynamics of natural enemies and plant populations as well as that of the pest population. In the example given in figure 1, the system consists of a predator, a pest insect, and the crop plant. Typically, each population is divided into a number of categories. Plants are described by variables that predict the weights of fruits, leaves, stems, and roots. Insect pests, parasites, and predators are usually divided into age classes.

While a principal purpose of such modeling is to gain understanding and thus direct research that may improve the accuracy of prediction, long-term prediction is not likely to be accurate if weather is a key factor. A number of studies in the project have illustrated the use of simulation modeling and also incorporated economic considerations (4-11).

Shoemaker (12) noted that the difficulty with the simulation method is that the simulation model must be calculated for each value of the pest control option in order to estimate the size of yields and overwintering populations, and as there may be thousands of possible combinations of pest control options, the computer costs of simulation would not be feasible. As a result, there has been considerable interest in the use of optimization methods, numerical methods that calculate the optimal strategy without recalculating

	Subclass	Numerical Value Stored on day t	Value Stored on day t+1
Parasite or Predator	egg	X_1	X_1
	larvae	X_2	X_2
	pupae	X_3	X_3
	adult	X_4	X_4
Pest Insect	egg	X_5	X_5
	1st instar	X_6	X_6
	2nd instar	X_7	X_7
	3rd instar	X_8	X_8
	4th instar	X_9	X_9
	pupae	X_{10}	X_{10}
	adult	X_{11}	X_{11}
Plant	fruit	X_{12}	X_{12}
	leaves	X_{13}	X_{13}
	stems	X_{14}	X_{14}
	roots	X_{15}	X_{15}
Weather inputs	temperature (T) solar radiation (S) precipitation (P)		temperature (T) solar radiation (S) precipitation (P)

Fig. 1. The structure of a typical pest management model.

the simulation model for each possible alternative. Two examples of the use of optimization methods in pest management modeling are the papers by Shoemaker (13) and Regev et al. (14). The use of optimization as well as simulation methods is discussed in a review paper by Shoemaker (12):

> Simulation and optimization models each have their advantages and disadvantages. Simulation models can include a large number of variables to give a detailed description of an ecosystem, but are very inefficient at evaluating a large number of management alternatives. Optimization methods, on the other hand, can efficiently

evaluate a large number of pest management options but only for systems which can be described by relatively few variables or by a special form of equations.

Unfortunately, most pest management problems require a large number of variables to describe the interactions in the system and they also present a large number of possible options. In these cases, the best procedure is to first develop a simulation model which describes the system and which has reasonably good predictive power. The second step is to develop a somewhat simpler version of the simulation model. The policies calculated to be optimal can then be substituted into the full-scale simulation model to see if they appear to be effective, economical policies when tested against a more detailed description of the system. The final test, of course, is an implementation in the field.

Perhaps the greatest benefit from the modeling effort has been that it has served as a focus of interest and has functioned to obtain the collaborative activities of the broad spectrum of biological and technological expertise that is needed (above). Thus, not only have entomologists and ecologists joined the program, but also plant physiologists, plant breeders, plant pathologists, economists, mathematicians, and engineers, as well as extension and private sector implementation people.

Using an ecosystem approach to pest management does not eliminate the need for insecticides. However, we do need to use insecticides much more judiciously and in ways that serve special purposes rather than for broad-spectrum effects sought by regularized or calendar date applications irrespective of whether or not populations threaten to exceed economic injury levels. We need to consider all the components of a philosophy that looks at the problem as one in applied ecology in order to determine the most judicious use of insecticides. Systems analysis is a way of organizing the effort and arriving at a synthesis of results.

BRIEF RÉSUMÉ OF SOME OF THE MAJOR ACCOMPLISHMENTS NOT REPORTED BY OTHER SPEAKERS

Cotton

Phillips et al. (15) pointed to the advances in systems analysis technology as applied to cotton pest management [see (6)], and described the results obtained concerning improved systems for management of insect pests of cotton. Cotton is

a highly plastic crop—that is, it exhibits drastic, rapid
responses to weather, pest damage, and management. The
responses may enhance yields or depress them [see (15)].

The program has verified earlier claims that use of insecticides has been to a large extent self-defeating, leading
to worse problems than those they used to alleviate. The
authors report that this IPM project has helped to develop a
systematic approach to agricultural research in general, an
approach that may lead to major contributions. Some of the
benefits can already be foreseen. The research undertaken
has been in four main areas: (1) host-plant resistance and
biological control; (2) basic and applied studies on the
factors influencing development of the cotton crop; (3) basic
studies to elucidate the physiological, behavioral, cultural,
and other processes affecting cotton insect populations; and
(4) use of these results in both traditional and systems modeling efforts to develop and upgrade pest control decision-
making technologies for implementation at the farmer level.
The latter has required a major research effort to develop
simulation and optimization models (above) for growth of the
crop, dynamics of the major insect pests, and the economic
impact of pests on cotton yield and quality.

Plant Breeding

Prior to this IPM program, there have been programs in
Mississippi, Texas, California, and elsewhere to develop cotton varieties resistant to pest insects and diseases. The
cooperative screening for resistance to boll weevil, plant
bugs, bollworms, aphids, mites, and whiteflies, continuing
from earlier USDA and Mississippi Agricultural Experiment
Station work, has been further developed. The Texas effort,
with USDA collaboration, is working with boll weevil, flea-
hopper, cotton diseases, and nematodes. In California similar collaborative efforts with USDA are seeking cottons resistant to lygus bugs, spider mites, nematodes, and verticillium. Beck and Maxwell (16) summarized these results up
to 1974, and listed a number of characters that confer resistance to particular insects. They also noted that certain
of these characters confer greater susceptibility to certain
other insect pests. For example, a cotton bearing the
glabrous, frego bract or nectariless character is more resistant to boll weevil and *Heliothis* but is more susceptible
to fleahoppers. The glandless (low gossypol), frego bract
and nectariless characters have been incorporated into short-
season cottons and are being tested for production and quality. The program has made real progress in developing and
establishing the potential value of growing short-season,

dwarf cottons, which avoid much of the boll weevil problem and
the *Heliothis* problem induced by use of treatments for boll
weevil, using some of the above characters.

Biological Control

Biological control in annual crops has been neglected,
although van den Bosch et al. (17) discusses a number of examples where natural enemies in cotton and other annual crops
have been effective. Falcon et al. (18) showed that use of
pesticides for *Lygus* control in California cotton induced
severe outbreaks of several secondary insect and mite pests.
This work was verified by Ehler et al. (19), Eveleens et al.
(20) and Gutierrez et al. (21). This project has elucidated
the reasons for these outbreaks, and the behavior and role
of each predator species has been quantified (e.g., searching
rates, host preferences, parts of plant inhabited, consumption
rates, etc.).

Three species of braconid parasite that attack tarnished plant bugs have been introduced in Mississippi and are
being evaluated. Studies in Texas revealed that a resident
parasite, *Bracon mellitor,* contributes significantly to boll
weevil mortality, and a survey of boll weevil or boll weevil-
associated parasites on a worldwide basis suggests that previously unknown boll weevil parasites may exist and might
furnish a significant tactic for integration into the overall
pest management system for cotton.

Practical Tests of Improved Pest Management Systems

A community-wide project for IPM of *Heliothis,* the key
pest of cotton, in a 50-square-mile area of the lower Mississippi flood plain in Arkansas was conducted in 1975 and
1976. The program was based on withholding use of insecticides unless a short-term model simulator predicted damage
from *Heliothis zea.* In 1976, 8 of the usual 10 almost weekly
treatments were predicted to be unnecessary and this proved
to be so. Thus, costs were correspondingly reduced, as were
the undesirable environmental effects of overuse of insecticides. The results also suggested that while insecticides
can be used to control bollworm, a mismanaged program is
almost as ineffective as no program at all.

The program in California started out from earlier fine
work of L.A. Falcon and coworkers in which the key pest status
of *Lygus* had been established, with secondary pest outbreaks
resulting from heavy treatments for *Lygus* control. Also, the
economic importance of assessing lygus bug populations relative to fruiting capacity of the plant was fully recognized.

Under the leadership of A.P. Gutierrez, this project has become much more quantitative. He and coworkers have (1) established the missing factor of the effect of lygus bug feeding, by sex and age class, on the shedding of young squares;
(2) developed a model of development of a field population of
cotton; (3) developed models for the population dynamics and
impact of several major insect pests of cotton, and for *Verticillium;* and (4) coupled these various components into a
model for decision making on cotton pest management [see (6,
15)]. Simplified recommendations are being developed for
computerized on-line delivery to field consultants, extension
specialists, and the farmer. In addition to the essential
contributions by entomologists in the project, the overall
progress has been possible only because of the additional
collaborative participation of researchers from many other
disciplines, including J.E. DeVay in plant pathology, Y. Wang
in mathematics, and U. Regev in agricultural economics.

Several practical experiments have been conducted in
Mississippi and Texas, which suggest that rather large potential benefits may accrue from adoption of IPM methods. The
Mississippi study suggested a cost of $16 to $31/acre, or
about half, under a pest monitored system (utilizing economic
injury potential for deciding upon a treatment) versus conventional treatments (22).

A most exciting event has been the development in Texas
of short-season, dwarf types of determinate fruiting cottons.
These cottons offer promise in greatly reducing insecticide
use, alleviating secondary pest outbreaks, using less water,
less fossil fuel, energy and labor, and less growing time.
A shorter growing time suggests that an extra crop per year
could be grown on the same land. Using less water and fossil
fuel suggests that the crop can be grown more cheaply and
profitably, while conserving these resources. Tests indicate that some of these cottons grown under more narrow
spacing produce even higher yields than conventional varieties
grown in conventional spacing (15).

Citrus

The United States produces about 36% of the world's
oranges, 30% of its lemons, and 7.5% of its grapefruits. U.S.
citizens consume about 42 lb of citrus per person per year.
In 1968-69 the national crop was valued at about $665 million.
In Florida, citrus comprised about 34% of the total value of
the state's agricultural commodities.

To accomplish this phenomenal production in the few
states that grow citrus, a heavily concentrated industry in

semiurban as well as in rural areas has developed, mainly in Florida, California, Arizona, and Texas. The crop has received intensive use of pesticides in recent years—in 1966 some 10 million lb of insecticides and 18 million lb of fungicides. The effects of such use more intimately concern the urban and semiurban population than insecticide use for a typical rural crop, such as cotton. This IPM project was launched on the assumption that the intensive use of insecticide could be readily reduced if a systematic effort were made to foster more biological control and ascertain the true need for use of insecticides in the major citrus-growing areas. The extensive experience with and studies on biological control in California, and to a less extent in Florida and Texas, justified this presumption. Little effort was thus made initially to adopt a program designed to utilize systems analysis and modeling, or any major tactic other than biological control and selective use of pesticides to protect the natural enemies.

With time, it became clear that modeling of citrus tree phenology, close monitoring of key citrus insect pests, and obtaining more fundamental information on the dynamics of certain key pest species would greatly help in the development of improved pest control systems. Economic analysis concerning both real and "cosmetic" effects of key citrus pests have been important. Ecosystem study plots designed to test the possibilities of using modified spray programs in Florida and California, in particular, were established (23).

In the Woodcrest plots in southern California it was found that *Aphytis melinus,* a principal parasite of the key pest, California red scale, would normally keep this pest under control in that area if seriously disturbing pesticides were not used. The data on yield and quality under several control regimes, including one on-line determined integrated control regime, suggest that in that area citrus of very high market quality can be produced using IPM programs that employ pesticides only minimally or, in some seasons or areas, not at all. An effective system is said to be ready for adoption on 76,000 acres of oranges and grapefruits in southern California and is probably adaptable for use on 45,000 acres of lemons[1].

On the other hand, in San Joaquin Valley citrus areas, *A. melinus* is ineffective and chemicals must be used as the major means of control of this key pest. Yet much more has been learned about red scale population dynamics, the impact of other natural enemies and the weather on the scale, about ways that insecticides can be used without completely destroying

[1] *L.A. Riehl, personal communication*

natural enemies (of other pests as well as red scale), and in
particular, how to monitor this scale's abundance in order to
time insecticide applications more critically, and to determine whether a treatment is even needed during a given season.

In Florida, the spectacular success of introduction of
the parasite *A. lingnanensis* for control of snow scale, *Un-
aspis citri,* has overshadowed the essential progress being
made toward better understanding of the basic features and
phenology of citrus tree growth and fruit production, and the
significant development (even before this project) of a system of integrated control for the complex of citrus pests, including citrus rust mite. It is becoming clear, however, that
if the introduction of this parasite had not been successful,
the problem of snow scale control by insecticides would have
required significant modification of integrated control programs and annual expenditure of some $8-10 million more for
insecticidal control for this pest alone (23).

Pines (Bark Beetles)

Bark beetles were considered so important in pine forests
that all of the project's effort has been centered on them.
They are the most destructive insect pest of pines, and nearly
every major U.S. pine type and region has a major bark beetle
problem. Waters (24) noted, ". . . All claims to the contrary,
we do not yet have a true management system for any major
forest insect, including pine bark beetles, in the United
States—operational surveys and control projects, yes—planned,
continuing management systems, no."

Most insect pests of agricultural crops and forests are
highly mobile, and their numbers and damage vary greatly in
space and time. Their seasonal activity patterns and relative
damage rate functions also differ greatly from place to place
and time to time. To manage such pests it is essential that
we develop the capability to anticipate and forecast their
occurrence, abundance, population trends, and potential damage
(25).

Our project on pine bark beetles has had three separate
subprojects, for the mountain pine beetle, *Dendroctonus ponder-
osae;* the western pine beetle, *D. brevicomis;* and the southern pine beetle, *D. frontalis.* The latter program was incorporated into the larger southern pine beetle program of the
USDA in 1974. For present purposes, I will review only the
program for the western pine beetle (WPB), and mainly its
pheromone program, as summarized by D.L. Wood. This program
was organized into four interconnected components, each of
which was designed to produce basic information for determining

the impact of selected treatment tactics or strategies on specified forest parameters and for using this information in benefit/cost models (25). These components are: (1) WPB population dynamics, (2) ponderosa pine stand dynamics, (3) impact, and (4) treatments. A considerable body of information has been developed on all of these components.

Here, we will deal only with progress achieved in the treatments component where attempts are made to lower WPB populations, which in turn lowers WPB-caused tree mortality. This creates a reduced impact and a subsequent benefit to input into stand dynamics and management models, along with the costs (including environmental costs) of the treatment(s).

Bedard and Wood (26) summarized some research with pheromones as follows:

> Pheromones of bark beetles release an attractant response that results in the location and colonization of trees, where the insects feed, mate and oviposit. In tree-killing species, such as the western pine beetle (*Dendroctonus brevicomis* LeConte) these attractant compounds elicit a synchronized invasion by several thousand beetles (27). The female beetles innoculate the tree with pathogenic fungi (28) that rapidly inactivate natural host resistance mechanisms. As a result, the tree dies and its tissues are exploited by adult beetles and their progeny. Beetles that fail to aggregate successfully do not reproduce. Thus attractant compounds derived from natural substrates have great potential for pest suppression. No insect suppression methods using pheromones are now registered for use, but many programs are under way to develop such methods. . . .
>
> . . . Regardless of the specific technique used, i.e., "trap-out", "baited tree", or "interruption", intensive evaluation of the treatment effects over large areas will be necessary. Methods of population and damage assessment must be available to make these evaluations of efficacy. Such evaluations are difficult and costly, and require the efforts of many individuals with a diversity of skills.

Two population manipulation methods, trap-out and interruption have been attempted for WPB (26,29).

Trap-out

Insects are attracted to traps baited with attractive compounds, and killed. The effects of two simultaneous trap-out treatments in one 65 sq. km area for five generations of

WPB were studied. Methods to determine the location of trees
killed by WPB and to estimate its within-tree populations and
natural enemies were known before we began the test, but had
never been applied to the evaluation of a pest suppression
program. Consequently, considerable modification and innovation was required. The number and location of killed trees
were estimated by using sequential aerial photography and
then subsequently verified on the ground. The distribution
and abundance of *D. brevicomis* and its principal natural
enemies were estimated by using a combination of aerial photo-
graphs and intensive within-tree sampling at three periods
during brood development. A suitable trap was designed that
provided a maximum trapping surface. The stability of each
pheromone compound was evaluated under field conditions. A
sequential sampling method was developed for estimating the
number of *D. brevicomis* and predators caught on suppression
traps at the end of the treatment period. Data management
specialists worked with all investigators to design forms and
protocols for efficient and accurate data capture, input,
file management, and analysis techniques. In all, nearly one
million WPBs were caught, and tree mortality declined
throughout the area.

Interruption

     Mass aggregation is reduced below the threshold number
required to kill the tree (27), either by increasing the concentration of attractant in the area or by introducing a
compound(s) that causes a reduced catch at pheromone-baited
traps. The team has demonstrated that release of three
pheromone components from many sites in a 0.81 hectare area
greatly lowers the catch of WPB on traps baited with the same
compounds in the center of the treated area[2].
     In addition to the suppression attempts, a survey of WPB
populations has been attempted by using attractant-baited
sticky traps (26,29). Analysis of site features associated
with the catch showed that (1) total catch was directly related to current tree mortality, (2) three peaks of abundance
were revealed during the season (these peaks coincided with
three distinct generations as determined by examination of
within-tree WPB development), and (3) slope and distance to
the nearest ponderosa pine were negatively correlated with
catch.
     In another study, by Gustafson et al. (30), the survey
trap catch data indicated a distinct pattern of spatial and

[2]*W.D. Bedard, unpublished data*

numerical distributions over the 130 sq. km area. The catch
at certain trap sites was consistently higher than at other
sites in each of the two trapping seasons. Also, the relative proportions of the total seasonal catch of the WPB and
two of its principal predators were in the same order of
magnitude between years.

The results of this research give us a better understanding of the role of pheromones in the population dynamics of
the WPB (31), and at the same time provide necessary information and methodology to register these compounds with EPA
for the suppression of this destructive bark beetle (29).
Discussions with the EPA have been initiated regarding requirements for such registration.

FUTURE PROSPECTS AND CONCLUSIONS

You have heard accounts of how insecticide usage may be
greatly reduced in the control of pests of apples, particularly by employment of pheromones for monitoring codling moth
and use of predators for mite control. You have heard how
plant breeding and cultural procedures such as use of a trap
crop may become very real components in an integrated control
program for soybean insect control and how proper integration
of research by entomologists and plant pathologists, for example, may avoid the adverse consequences of each discipline
proceeding alone. This program has, we feel, gone very far
toward saving the soybean industry in the South from conversion to a pattern of insecticide usage that may well have put
soybean insect control in the desperate situation that cotton
was in a few years ago. Coincident with these gains comes an
increased profit, reduced adverse environmental consequences,
and reduced use of fossil fuels.

We are on the verge of transforming insect control from a
system of half science and half guesses to one based primarily on facts, in which the promotion of insecticides will no
longer be a decisive determinant of what is to be done. In
doing so we are also entering the era when not only insect
control but all pest control and indeed crop production itself will be more scientifically based. Priorities will be
determined through an orderly process of farm decision making,
based on actual results from monitoring the fields for the
conditions that affect crop growth and yield.

A corps of highly trained professionals will be needed to
monitor the major features required. A weather network designed and computerized to satisfy the needs for modeling
events throughout the nation is needed. You have heard how
such a network is effectively used in pest management in

Michigan. Nationwide, such a network could serve many other
uses as well. You have heard how a telecommunication network, tied into a data bank of pest incidence, crop conditions, and pest control tactics can be used to update our
traditional extension service. Without such updating, the
extension service could not begin to cope with future needs.
Private consultants, too, will be able to utilize the new
pest control guidelines and develop their own monitoring
systems to put into mini-computers, which will include form-
ulae for optimizing their decisions on pest control.

We believe that this Symposium has demonstrated the value
of a new approach to research and application technology in
pest control. This is illustrated by the gains that have been
achieved by use of better information concerning: (1) "need
for treatment" criteria; (2) "time to treat" information;
(3) effects of given pest species at different densities and
stages of development, and for different stages of crop development, on quantity and quality of yield; (4) capacity of
the crop to produce yield, and the relationship with weather,
soil, agronomic, and pest complex factors (often involving
compensation phenomena); and (5) the practical, economical,
and environmentally promising results from field tests using
modified pest management systems based on the strategy of
pest containment and using systems analysis.

Some of the practical benefits you have heard have been
gained, and others could be gained, without using systems
analysis and modeling. Other gains have been made or contributed to because of the use of the systems analysis approach. The systems approach is, in fact, almost synonymous
with the first dictum of integrated control, "consider the
(whole) ecosystem." We feel that the tools of systems analysis offer us a path by which we can orderly establish the
research needs, explore the biological, physical, economic,
and social problems that are suggested, and then assess the
results as components of a single interlocked system. Needed
are facts and more facts, rather than "educated guesses,"
however high the level of experience and sage the head of the
one who would recommend on a basis of less than facts. On
the other hand, we are the first to admit that only a small
proportion of the total sophistication necessary in research
will be practical for farm-by-farm use in daily decision
making. But it is only by developing an understanding in
depth that we can confidently settle on the main criteria,
neglecting endless details, and simplify the monitoring and
delivery system, as must be done if we are to establish
realistic, implementable IPM programs on a crop-wide national
scale.

We wish now to insert our personal opinion, not

necessarily concurred in by a majority, or even any other program participants. The IBP/NSF/EPA program had perhaps an "unlikely" beginning, in which a series of national movements and political events as well as basic scientific needs had a great deal to do with the program's acceptance and funding. These events embraced such things as the population explosion and the great demand for food, the lessening effectiveness of insecticides because of resistance problems, the global concern for a clean environment and preservation of endangered species, human health hazards posed by insecticides, and lastly, the ban of DDT. In any event, the project was funded and has enjoyed continued support.

We would point to three factors that have hindered development and achievement of improved pest control, with maximum benefits for society. The *first* is that the chemical industry has for too long dominated the pest control scene, and this has resulted in an almost complete departure from some of the older more ecologically based methods of pest control. A virtual army of insecticide salesmen have in some parts of the country practically replaced the traditional dependence of the farmer on his university, its researchers, and extension advisors for advice. There must be some way that this can be corrected. We should put a force of professional entomologists in the field to do the necessary monitoring and assessment of the need for treatment and to ascertain what measures, if any, are best. Their objectivity should be preserved by their independence from chemical companies. A highly efficient pest control advising system is needed, and support in its implementation may be indicated.

The *second* is that the method of funding and managing research programs to develop improved pest control, i.e., IPM, must allow for some changes. Existing routes of funding are through small individual research grants on small pieces of basic or applied research, or through the USDA. USDA funding either (through ARS) fails to bring in a broad-based vigorous input from research scientists in the universities, or when it does so (through CSRS), channels money to state experiment station directors, deans, heads of departments, etc., by "formula funding," such that the money getting to researchers is usually a little here and a little there. It is thus very difficult, if not impossible, to fund a major cropbelt-wide team project (i.e., a coordinated IPM program) with the strength required for success.

Thirdly, management and research priorities must be revamped. At present, most management of pest control research is automatically subject to the cross-currents, opposing viewpoints, and yes, parochialisms or special backgrounds of the administrators at different levels in the several

universities usually involved in coordinated programs that
now exist. A program of appropriate scope and technical
depth, centered on use of systems science and modeling, as a
means of setting research priorities, guiding research, evaluating results, and optimizing economic and social benefits
to the farmer and society requires a strong centralized management largely independent of domination by these administrators, and lacking the dilution of dollars as they filter
down to various individuals.

Finally, this large IPM program supported jointly by NSF
and EPA, in which USDA has assisted, with the help of a
broadly based Executive Committee and Steering Committee,
became possible because government realized the need for such
a *centrally managed* and *block-funded* effort. The program
that we and others have described attests to the success that
can be had when such programs are solidly established and
strongly supported. But we have just begun! We need to
establish more solidly the insect control programs we envis-
age for the six crops we have worked with, and develop
similar programs for all our crops, and to look at the livestock pest and urban situations. But we need first to bring
in the other kinds of pests—plant pathogens, nematodes, and
weeds (which we have not done), and the whole gamut of crop
and livestock production. A farming operation is a complex
system. By using systems science we can serve the farmer
better than we have. The farmer deserves more than he has
gotten in the past and more than the most dedicated individual scientists can give him. He needs to have the whole farm
operation looked at as a unit, his options organized, and the
consequences detailed for him. Moreover, if the family farmer is going to be able to meet the competition from the ever-
increasing corporation operation, he will need the clear insight and predictive potential for cost/benefit analysis and
decision making that systems science and effective information afford.

REFERENCES

1. Huffaker, C.B. 1972. Ecological management of pest systems. pp. 313-42 *in* Challenging Biological Problems.
   J.A. Behnke, ed. Oxford University Press, New York.
2. Stern, V.M., R.F. Smith, R. van den Bosch, K.S. Hagen.
   1959. The integration of chemical and biological control
   of the spotted alfalfa aphid. The integrated control concept. Hilgardia 29:81-101.
3. Huffaker, C.B. and R.F. Smith. In ms. Rationale, organization and development of a national integrated pest

management project. chapter 1 *in* New Technology of Pest Control. C.B. Huffaker, ed.

4. Asquith, D., B.A. Croft, S.C. Hoyt, E.H. Glass, R.E. Rice. In ms. Systems approach and general accomplishments toward better insect control in pome and stone fruits. chapter 9 *in* New Technology of Pest Control. C.B. Huffaker, ed.

5. Coulson, R.N., W.A. Leuschner, J.L. Foltz, P.E. Pulley, F.P. Hain, T.L. Payne. In ms. Approach to research and forest management for southern pine beetle control. chapter 14 *in* New Technology of Pest Control. C.B. Huffaker, ed.

6. Gutierrez, A.P., Y. Wang, D.W. DeMichele, R. Skeith, L.G. Brown. In ms. Systems approach to research and decision-making for cotton pest control. chapter 6 *in* New Technology of Pest Control. C.B. Huffaker, ed.

7. Luck, R.F., J.C. Allen, C. McCoy. In ms. Systems approach to research and decision-making for citrus pest control. chapter 11 *in* New Technology of Pest Control. C.B. Huffaker, ed.

8. Rudd, W.G., W.G. Ruesink, L.D. Newsom, D.C. Herzog, R.L. Jensen, N.F. Marsolan. In ms. Systems approach to research and decision-making for soybean pest control. chapter 4 *in* New Technology of Pest Control. C.B. Huffaker, ed.

9. Ruesink, W.G., C.A. Shoemaker, A.P. Gutierrez, G.W. Fick. In ms. Systems approach to research and decision-making for alfalfa pest control. chapter 8 *in* New Technology of Pest Control. C.B. Huffaker, ed.

10. Stark, R.W. and Collaborators. In ms. Approach to research and forest management for mountain pine beetle control. chapter 12 *in* New Technology of Pest Control. C.B. Huffaker, ed.

11. Wood, D.L. and Collaborators. In ms. Approach to research and forest management for western pine beetle control. chapter 13 *in* New Technology of Pest Control. C.B. Huffaker, ed.

12. Shoemaker, C.A. In ms. The role of systems science and modeling. chapter 2 *in* New Technology of Pest Control. C.B. Huffaker, ed.

13. Shoemaker, C.A. 1977. Pest management models of crop ecosystems. pp. 545-74 *in* Ecosystem Modelling in Theory and Practice. C.A. Hall and J. Day, eds. Wiley-Interscience, New York.

14. Regev, U., A.P. Gutierrez, G. Feder. 1976. Pest as a common property resource: a case study in the control of the alfalfa weevil. Am. J. Agr. Econ. 58:187-97.

15. Phillips, J.R., A.P. Gutierrez, P.L. Adkisson. In ms. General accomplishments toward better insect control in cotton. chapter 5 *in* New Technology of Pest Control.

C.B. Huffaker, ed.

16. Beck, S.D. and F.G. Maxwell. 1976. Use of plant resis-
tance. pp. 615-36 *in* Theory and Practice of Biological
Control. C.B. Huffaker and P.S. Messenger, eds. Academic
Press, New York.

17. van den Bosch, R., O. Beingolea, M. Hafez, L.A. Falcon.
1976. Biological control of insect pests of row crops. pp.
443-56 *in* Theory and Practice of Biological Control. C.B.
Huffaker and P.S. Messenger, eds. Academic Press, New
York.

18. Falcon, L.A., R. van den Bosch, C.A. Ferris, L.K. Strom-
berg, L.K. Etzel, R.E. Stinner, T.F. Leigh. 1968. A comparison of season-long cotton-pest-control programs in
California during 1966. J. Econ. Entomol. 61:633-42.

19. Ehler, L.E., K.G. Eveleens, R. van den Bosch. 1973. An
evaluation of some natural enemies of cabbage looper in
cotton in California. Environ. Entomol. 2:1009-15.

20. Eveleens, K.G., R. van den Bosch, L.E. Ehler. 1973. Secondary outbreak induction of beet armyworm by experimental
insecticide applications in cotton in California. Environ.
Entomol. 2:497-503.

21. Gutierrez, A.P., L.A. Falcon, W. Loew, P.A. Leipzig, R.
van den Bosch. 1975. An analysis of cotton production in
California: a model for Acala cotton and the effects of
defoliators on its yields. Environ. Entomol. 4:125-36.

22. Smith, R.F., C.B. Huffaker, P.L. Adkisson, L.D. Newsom.
1974. Progress achieved in the implementation of integrated control projects in the USA and tropical countries.
EPPO Bull. 4:221-39.

23. Riehl, L.A., R.F. Brooks, T.W. Fisher, C. McCoy. In ms.
General accomplishments toward improving integrated pest
management for citrus. chapter 10 *in* New Technology of
Pest Control. C.B. Huffaker, ed.

24. Waters, W.E. 1974. Systems approach to managing pine bark
beetles. pp. 12-14 *in* Southern Pine Beetle Symposium. T.
Payne, R. Coulson, R. Thatcher, eds. Texas Agricultural
Experiment Station and U.S. Forest Service.

25. Waters, W.E. and B. Ewing. 1976. Development and role of
predictive modeling in pest management systems—insects.
pp. 19-27 *in* Modeling for Pest Management. R. Tummala, D.
Haynes, B. Croft. eds. Michigan State University, East
Lansing.

26. Bedard, W.D. and D.L. Wood. 1974. Management of pine bark
beetles—a case history of the western pine beetle, pp.
15-20 *in* Southern Pine Beetle Symposium. T. Payne, R.
Coulson, R. Thatcher, eds. Texas Agricultural Experiment
Station and U.S. Forest Service.

27. Wood, D.L. 1972. Selection and colonization of ponderosa pine bark beetles. *in* Insect/Plant Relationships. H.F. van Emden, ed. Symp. Roy. Entomol. Soc. (London) 6:101-17.
28. Whitney, H.S. and F.W. Cobb, Jr. 1972. Non-staining fungi associated with the bark beetle *Dendroctonus brevicomis* (Coleoptera: Scolytidae) on *Pinus ponderosa*. Can. J. Bot. 50:1943-45.
29. Bedard, W.D., D.L. Wood, P.L. Tilden. In ms. Use of behavior-modifying chemicals to lower western pine beetle-caused tree mortality and to protect trees.
30. Gustafson, R.W., W.D. Bedard, D.L. Wood. 1971. Field evaluation of synthetic pheromones for the suppression and survey of the western pine beetle. McCloud Flats, Shasta—Trinity National Forest. U.S. Forest Service, San Francisco, California.
31. Wood, D.L. and W.D. Bedard. 1977. The role of pheromones in the population dynamics of the western pine beetle. pp. 643-52 *in* Proc. XV Int. Congr. Entomol., Entomol. Soc. Am., College Park, Maryland.

DISCUSSION

R. RILEY: You are implying that considerable resources will be required to put IPM into crop production. About how much, or what percentage of the current resource base will be required?

C. HUFFAKER: In dollars, I don't know. We now have the tools to look at the entire farming situation, with a lot of information on plant modeling, for example. Our program has been concerned only with insect and plant modeling; we did not include plant pathology, nematodes, or weeds.

T. CLARKE: Should soil quality be included?

C. HUFFAKER: Yes; everything in the farm operation, including fertilization and irrigation should be included.

R.D. O'BRIEN: There is little question that this is a superior program with respect to minimizing pesticide use, environmental problems, etc. and is probably cost-effective in terms of reducing straightforward costs of pesticide use; but what about total costs, including the increased cost involved in managerial skills?

C. HUFFAKER: We have not attempted to determine the full costs of using pesticides, including external costs. Our principal objective has been to develop more rational and better long-term programs. Our only cost analysis has been based on profits to the growers.

POLICY COHERENCE THROUGH A REDEFINITION OF THE PEST CONTROL
PROBLEM, OR "IF YOU CAN'T BEAT 'EM, JOIN 'EM"

Thomas P. Grumbly[1]

Special Assistant to the Commissioner
Food and Drug Administration
Rockville, Maryland

INTRODUCTION

In the fall of 1975 a funny thing happened on the way to
a coherent pest control policy in the United States. The
House of Representatives barely defeated a proposed amendment
to the Federal Insecticide, Fungicide and Rodenticide Act
(FIFRA), which would have given the Secretary of Agriculture
concurrent jurisdiction with the Administrator of EPA over
pesticide regulation. In effect, this would have given veto
power over pesticides back to USDA. The closeness of the vote
shocked many in the environmental community and was sympto-
matic of an emotional and continuing struggle over the direction of pest control policy.

That struggle, more complicated than some would think,
keeps the federal government from having a set of rational
policies in the field of pest control. To be sure, we have
laws that purport to outline policy direction by providing
regulatory authorities. We have programs and processes designed to weigh the risks and benefits of particular substances (assuming we have the technical capability to do so).
These programs, however, are only palliatives designed to
treat symptoms of more fundamental political problems.

This paper constitutes.an attempt to begin plumbing the
political and analytical depths of the pest control problem

[1]The author was formerly Examiner for Food and Agriculture, U.S. Office of Management and Budget, Executive Office of
the President. These views are his own and do not necessarily
reflect the policy of the United States Government.

261

of which pesticides is a major part, define "it" in a way that
is manageable from the perspective of either the Science
Advisor or the President, outline some alternative courses of
action, and, finally, make some preliminary recommendations.
In the course of such discussion, obstacles and incentives to
the implementation of sophisticated pest management practices
ought to become quite evident.

PLACING THE PEST CONTROL PROBLEM

Fifteen years after the publication of *Silent Spring,* we
still place enormous sums of pesticide on the land. After
nearly ten years of concentrated discussion surrounding the
use of integrated pest management, the government still talks
blithely of totally eliminating damaging insects such as the
screwworm or the boll weevil. After numerous reports that
port and border inspection does little or nothing to keep
dangerous insects out of the United States, we persist in
spending large sums of money on these activities.

A superficial response to the persistence of these activities is to label chemical companies, agribusiness firms, and
government as villainous, shortsighted, and stupid. Another
superficial response, in my judgment, is to assume that the
companies have an enormous "special interest" hold over the
government, which needs to be counteracted by equally strong
"public interest" action from a different ideological stance.
Each of these "hypotheses" has elements of truth in it, but
each tends to overestimate the applicability of conspiracy
theory to political life. In view of our recent past, this
tendency may be understandable, but, in my judgment, it is no
less deficient.

These three manifestations of pest control problems—the
emphases on pesticides, eradication, and outmoded inspection
practices—stem, I believe, from some fairly strong philosophical incentives in the society—
1. a widespread belief among farmers, and indeed, among
   the public as a whole, that *maximum* food production,
   at the cheapest short-run cost, is a necessity to meet
   both domestic and international responsibilities;
2. a proper belief in simplicity on the part of almost
   everyone, and a particularly strong wariness on the
   part of bureaucrats to do anything too complicated
   (for a variety of reasons);
3. an almost overwhelming desire for the great technological fix—for the immediate application of science to
   solve problems in a definitive way.
Each of these aspects needs some explanation.

The Maximum Production Construct

Over the years, U.S. policy has consistently favored maximum production by sending price signals to farmers to plant in excess of demonstrated demand (indeed, even our famous "paying people not to grow" policy was merely a countermeasure to dampen our general policy thrust). Either through price support mechanisms or exhortation, we have encouraged full production to meet a vast array of real or imagined, domestic or international problems. The emphasis on quantity is enormous. The use of pesticides as pest control agents is quick and has been marvelously effective in providing enormous short-run productivity gains. Any attack on pesticides runs immediately into the argument that these chemicals are necessary to ensure full production. Eradication measures offer the hope that cost constraints on full productivity will be successfully eliminated at some fixed point in the not-too-distant future. Import inspection is fueled by the oversimplified spectre of ravenous new insects ravaging our production capacity and eating into our balance of payments.

The Drive for Simplicity

Common sense, short-run production functions, bureaucratic battle fatigue, and empire building all contribute an enormous desire to solve problems with single-pronged, goal-directed measures. Pesticides are easily applied, need little monitoring, and in the short run, at least, seem to work quite well. Secondary pest problems, long-term pollution effects, and genetic resistance are all aspects of pesticide application easily (too easily) placed into the class of "externalities" by farmers and their friends. The ability to either ignore such externalities or to regard them as someone else's responsibility enables many to maintain the fiction of simplicity.

In the wake of complex social programs, which seemed to perform imperfectly, many government planners have become wary of scientific schemes requiring the sucessful interaction of many parts for total program success. In a fairly adroit manner, advocates of various eradication programs have managed to make their tactics appear simple (e.g., dependence on the "sterile male" technique), while depicting pest management schemes as extraordinarily complex. Proponents of integrated pest management have not helped themselves by emphasizing the complex interactive nature of the strategies, and indeed by depicting many of those strategies as still being in the fundamental (read nonapplicable) research stage.

The truth is, of course, that eradication strategies may
be very complex, whereas management strategies can be relatively simple, although never-ending. Through better public
relations, however, eradication proponents have managed to
take the "simple" ground, a ground to which common sense
often drives us.

It must also be said, in all honesty, that the simplicity
of eradication programs is often boosted by bureaucratic
scientists who desire to ride such programs to agency promi-
nence. It is a well-known and time-tested fact that bureaucratic politics get in the way of science, and that some men
will bring dishonor to their disciplines by oversimplifying or
overassuming in presentations to nonscientific government
decision makers. Those in power, when faced with hundreds
of decisions, often crave for scientists who will speak to
them in assuring terms, rather than in the traditional "on
the one hand, but on the other hand" scientific consulting
role. Pesticide usage and eradication programs represent the
"simple" choice when confronted with alternative technical
solutions. Import inspection represents the "simple" choice
when no alternate feasible solutions are broached.

The Technological Fix

While related to the drive for "simplicity" expressed
above, the option of "technological fix" is more important in
that it crosscuts all of science policy. This is no news, but
the great medical and atomic energy breakthroughs of the not-
too-distant past still drive Americans, and hence their government, to opt for technology over alternative policies.

The use of pesticides revolutionized (so it is said)
American agriculture, and one suspects they will always have
their place. In the absence of overpowering and easily understood gross evidence that pesticides are harmful, it is difficult to see significant shifts from a proven "technological
fix" (assuming that the relative price of that technology
remained constant). Integrated pest management, to be adopted
on a large scale, will not only have to be economic, but will
have to provide its own technological razzle-dazzle or piggy-
back on some other concept of equal appeal to the American
farmer.

Eradication programs are the epitome of the technological
fix, depending as they usually do on the sucessful application
of one particular technology (e.g., sterilization of boll weevils or screwworms, the application of ethylene gas to
witchweed, the use of mirex on fire ants). It is again evident that the political drive for successful, final solutions

is encouraged by the "technological fix," particularly since
such "fixes" do away with hard political choices that may
involve coercion or the necessity to tell constituents that
some problems must be lived with.

DEFINING THE PEST CONTROL PROBLEM

In the wake of the preceding discussion, the question of
proper government pest control strategy becomes more complicated, but, ironically, more manageable in that we begin to
reach the roots of obstacles currently preventing us from
reducing pesticide use generally and implementing management
strategies on a broad scale. The problem changes from one of
convincing evil-minded bureaucrats and Congressmen to compel
the reduction of pesticide usage to either: (1) reducing
pesticide usage by tying into other deep-seated incentives
in the society that keep current strategies useful, or (2)
changing, at least on the margin, the prevailing societal
incentives.

This kind of redefinition will, I think, be helpful both
from a substantive policy and from a political perspective.
The former remains to be seen, but the latter is clear to the
extent that the albeit unsuccessful House vote represented the
beginning of a counterattack on "city-slicking," lawyerly,
"do-gooder" environmentalists who challenged the establishment
with a problem definition of the conspiratorial variety expressed earlier.

Before we proceed to manipulate solutions under either
Formulation A (Working with the System) or Formulation B
(Changing the System), we need to set forth the institutional
setting to bring some reality to our ethereal constructs.
Imagine a setting in which fledgling regulatory and scientific
agencies, with small initial Congressional constituencies and
little clout with the President, attempt to take on a hundred-
year-old Colossus of Rural America in which the leadership is
close to the President; moreover, the intellectual vanguard
of the Colossus has been isolated from and warring with intellectual underpinnings of the society's "big" science institutions for at least a generation. Compare this to the Environmental Protection Agency, National Science Foundation, Department of Agriculture, and the land-grant universities of the
early- and mid-1970s.

Resemblances are not coincidental. Assume that the power
alignment will not change by a quantum factor even with a new
and environmentally different administration. With this assumption, *then the locus of our discussion about pest control
must continue to revolve about USDA.* If, in fact, the power

alignment changes drastically, federal policy as dictated in
Washington may more clearly meet objectives of pesticide reduction. Unless the Department of Agriculture itself changes,
however, the enormous field apparatus of the Department and
the land-grant universities will continue to raise havoc with
the actual implementation of integrated pest management, non-
eradication objectives.

Consequently, effective policy formulation requires attention to the field structure of the Great Colossus, embodied
in the Agricultural Stabilization and Conservation Service,
the Soil Conservation Service, the state experiment stations,
their cousins in the Agricultural Research Service, and the
Extension Service.

IMPLEMENTING FORMULATION A—WORKING WITH THE SYSTEM

For the purposes of simplicity, let us assume that a
policy will be constructed that depends upon the continuation
of all three major incentives described above: (1) maximum
production push, (2) simplicity of solution, and (3) emphasis
on technological certainty.

Our objective is to reduce pesticide usage without turning the fields over to the insects, and to engage the support
of relevant USDA and other personnel in this task. A solution
might look like the following:

1. Permit marginally higher "income support rates" in
   return for adopting a specified pest management plan
   on the land; or raise the maximum payment level in
   return for improved management strategies.
2. Construct a technological menu that minimizes the
   number of pesticide applications, still uses pesticides in limited sums, and emphasizes long-run problem
   solution through "razzle-dazzle" strategies such as
   host-plant resistance.

This strategy minimizes the use of computers (viewed by
many as a complicating factor), maximizes on field scouts
(probably—much along the lines of present extension systems),
but holds out two technological fixes as possibilities: (1)
the known one of pesticides, and (2) the more unsure yet al-
luring possibility of host-plant resistance. One should, of
course, not be wedded to any particular form of "technological
fix" (under this formulation), but this solution suggests that
pest management research might be skewed toward the development of a *single* alternative technology to pesticides in order
to attract public support. It also suggests undertaking a
concentrated, well-publicized program on a particular technology.

This formulation probably entails additional substantial budget costs, through its ties to the more traditional farm programs. The optional amount is unclear, although some idea could be obtained through looks at the current externalities imposed upon the environment by pesticides, as well as the net-discounted benefits theoretically offered by eradication programs.

IMPLEMENTING FORMULATION B—CHANGING THE SYSTEM

"Changing the system," or altering the strong incentives mentioned previously, is different from "bucking the system." This strategy must, therefore, be seen in a larger time perspective than Formulation A and must have sufficient feedback loops to keep the ship "on course." Our objectives are relatively clear: (1) change our farm policy from one of maximum short-run production to one of maximizing long-run net benefits to Americans from our production system, (2) increase our ability to handle complex solutions to problems, and (3) move away from the notion that single infusions of massive technology (in the absence of other economic or organization policies) can solve our problems.

A suggested set of policies to accomplish these objectives might look like the following:

1. A gradual return to a competitive agricultural supply system in which the proper price incentives are sent to farmers, coupled with a modest resolve to dampen excessive consumer price fluctuation. We are seeing part of this battle being played out now (in June 1977) as the farm bill moves to the finale.

2. The maintenance of some government price and acreage controls so that incentives for pest management such as those in Formulation A can still be implemented.

3. A clarification by the President and the Congress of the realistic U.S. role in world food production.

4. Emphasis in our agricultural educational systems on the rudiments of quantitative analysis, as a basis for understanding things such as systems analysis and computer technology.

5. A broadly distributed research program by government, aimed not at a single technological solution to pest control problems, but directed at budget problems and behavioral change; in essence, the growth of public policy research at our land-grant universities, and the increased mingling of those universities with the other scientific and policy-making institutions in the society.

6. Strong, continuous, and systematic evaluation of the extension science education systems in agriculture by the leadership of their own universities.

This strategy demands that we have clear signals from the top of our government about the objectives of our overall agricultural policy. That is difficult enough for policy makers who prefer fuzziness. More fundamentally, however, it requires a gradual shift by our great agricultural institutions, the land-grant universities, to a position *vis-à-vis* government more like that enjoyed by many private institutions. We need questioning, critical scientists to challenge prevailing assumptions, to keep us from being too simple or too eager to accept the false prophets of the profession. The current relationship between USDA and the land-grant universities is too cozy to be healthy. Only the leadership of these great universities can alter this, and by leadership I mean the presidents of these institutions.

RECOMMENDATIONS AND CONCLUSION

As might be expected, it seems entirely feasible to combine many of the policies outlined under the two formulations. In the short run, it seems more likely that piggybacking onto existing incentives will be helpful a la Formulation A. As this "piggybacking" generates internal dynamics, the incentives already begin to shift on the margin. At that point, a strong effort along the lines of Formulation B becomes a greater possibility.

We have approached the pest control problem from the implicit assumption that present government policies, and particularly those of USDA, are inadequate. Although USDA is the prime mover in terms of attitudes and, hence, merits attention, other government institutions have often been heavy-handed and overly self-righteous. The House vote in 1975 reflected, perhaps, a healthy response from part of the society feeling genuinely oppressed by government. Unless new approaches are adopted utilizing and playing upon strong interests in the agricultural and, indeed, the entire American community, EPA and others in the environmental community can expect more well-organized resistances to innovative pest control schemes. The FIFRA vote in 1975 was a storm warning. The entire pest management community would do well to heed it.

DISCUSSION

L. BOWEN: I would like to point out that there are only two
farmers and two independent pest management consultants
present.

T.P. GRUMBLY: Yes; unfortunately, what we have been saying
has not been getting out from the academic and government community.

L. BOWEN: There are a number of extension people here who
might be able to disseminate this information.

B. DAY: People in the Extension Service are often opinionated
and independent, and are not likely simply to push a party
line from Washington.

T.P. GRUMBLY: Maybe we can convince them at least to be neu-
tral.

UNIDENTIFIED: If IPM programs are successful, extension agents
will be behind them.

T.P. GRUMBLY: The county agent is under a number of conflicting incentives, from growers, local governments, and the university. His activities are not always predictable.

T. CLARKE: The assumption of the county agent's influence may
be overemphasized; many farmers ignore his advice.

J. GOOD: People here have been working "within the system,"
but still have encountered much opposition and difficulty.

T.P. GRUMBLY: Dr. Cutler's comments indicated a change in
policy within the government, with perhaps more cooperation
between USDA and EPA. But whether that can be transmitted out
to the American public remains to be seen.

M. HINKLE: Concerning the 1975 vote, my interpretation of what
it meant was different from yours. I thought the vote was
close the first time because of the negligence of the bill's
opponents—no one dreamed that it could go through. The
second vote was not close. This year, another attempt was
made to pass something similar, and there was no chance of
its getting through.

T.P. GRUMBLY: Well, I viewed the unexpectedly close vote as a
storm-warning, especially since USDA didn't even lobby very
hard for the bill.

INDUSTRY PERSPECTIVES ON PEST MANAGEMENT

Richard R. Whetstone

San Ramon Business Centers
Shell Chemical Company
San Ramon, California

To me integrated pest management is common sense—the application of the best available techniques to control pests economically with acceptable risk to humans and the environment. To the farmer, it is a part—a key part—of total farm management. Pest management, therefore, must be in harmony with other factors of farm management and must change as those factors change. For example, the trend toward conservation or minimum tillage in response to regulation, under the Clean Water Act, will increase problems from weeds, insects, fungi, and nematodes and therefore increase the need for, and the usage of, pesticides or other means of pest control. The farmer wants and must have available well-established, reliable pest control techniques from which he can select those best suited to his needs. At present and at least to the year 2000, most farmers will rely on chemical pesticides supplied by the pesticide industry.

My purpose is to present a Research and Development manager's view of the barriers the industry faces in developing new pesticides, and to suggest changes. However, let's first look at the challenge for new pesticides, using examples from insect control because it is there that the opportunities and pressures for nonchemical methods of control are the greatest.

The following picture of insect control on cotton is based on the recent excellent review by Bottrell and Adkisson (1). In the late 1940s and early 1950s, highly effective insecticides, largely chlorinated insecticides, controlled all of the serious pest insects of cotton. These insecticides produced an impact perhaps unlike any other modern agricultural technological advancement. They enabled the cotton farmer to develop strategies for selection and control

of the various inputs to maximize crop returns. Average
yields of lint cotton per acre in 1946-1955 were 19% higher
than in the previous 10 years. In addition, the farmer was
able to grow cotton varieties with higher quality, longer
staple lint, which brought higher prices. This short Camelot
soon ended because of insect resistance, outbreaks of
secondary insect pests, and environmental problems. The use
of chlorinated insecticides against corn soil insects
followed a similar pattern.

In the United States, chlorinated insecticides may soon
join the dinosaurs as museum specimens. However, they and
related PCBs, PBBs, and dioxins have burned in the public
consciousness the image that synthetic organic compounds
generally are highly toxic, nonselective and residual—very
different from the compounds found in nature. Thirty year's
experience with pesticide research has given me a very
different view. In screening of many thousands of random
organic compounds, only about 5% have been significantly toxic
to plants, insects, fungi, and nematodes at dosages of 10 to
100 times that of commercial pesticides. Only a few of the
active compounds from these tests and from directed synthesis
are useful as pesticides. The most common cause of failure
is rapid breakdown in the organism or the environment. With
rare exceptions, the pesticide research problem is to increase
rather than decrease both biological activity and residual
life. For example, 25 years of research was required from the
first synthesis of analogs of the natural pyrethrins to build
in the activity and stability required for usefulness on
crops. The residual, toxic halogenated compounds mentioned
earlier stand out like NBA basketball players in a crowd.
I suspect that the distribution of toxicity and residual
properties in random synthetic organic compounds is not
greatly different from that of natural compounds.

The chlorinated insecticides were largely replaced for
control of cotton insect by more acutely toxic but less
residual phosphate and carbamate products. Now we have
highly effective, new synthetic pyrethroids. Another
insecticide of a new type, Dimilin, may for the first time
allow real hope for the eradication of the boll weevil. The
pesticide industry is meeting the need for new, safer
insecticides more compatible with the environment.

The insecticide chemist rapidly develops a profound
respect for the abilities of insects to adapt to pesticides
through modification of detoxification mechanisms, active
sites of critical enzymes, and membranes. The evidence
suggests that insects may be adaptable also to many types of
biological control. The screwworm problem has reappeared
because, unexpectedly, sterile males are not competitive with

normal males. I suggest that the development of reliable, effective biological insect control techniques to the point of confident usage by farmers may take as long as for a new class of insecticides, and may need replacement about as often. In addition, biological insect control methods are often slow or incomplete. Biological control methods are generally less developed for control of weeds, fungi, and nematodes, except for crop varieties resistant to fungi.

Production of food to feed the growing world population is clearly the primary world problem, both now and in the future. For this country, food and fiber production is also the best hope of paying for the increasing costs of energy imports. By conservative projections, the world needs to double total food production by the year 2000. Further, U.S. production should increase from the present 17% of the world food supply to between 20% and 25%. This requires that U.S. food production increase to more than $2\frac{1}{2}$ times the present level. Our farmers need to maximize the yield and quality of all crops as the cotton farmer did in 1946-1955. In order to do this, we must provide the farmer with the best pest control techniques possible, both chemical and biological. The federal govenment is the champion of biological pest control and the source of most of the financial support. However, it must not forget that chemical pesticides will continue to be needed, and that the pesticide industry is the only source of new, better, safer pesticides compatible with biological control methods.

Can and *will* the pesticide industry find and develop the new pesticides that will continue to be needed?

Let's first look at the technical problem. Certainly, the difficulty of finding significant new pesticides is increasing, as measured, for example, by the average number of compounds that must be made and tested to find a new product. However, this is counterbalanced in part by the increasing sophistication of industry research on new pesticides. As a research manager, I am confident that the pesticide industry can continue to find new, improved, and safer pesticides, including the new insecticides needed to stay ahead of insect resistance.

However, the pesticide industry relies on government and universities for basic research. Recently, much of the basic research has been shifted to biological pest control. In addition, the staff of USDA Agricultural Research Service (ARS) has shrunk in recent years, despite increased needs.

I make the following recommendations:
1. The USDA ARS staff should be increased, especially those involved in pest control.
2. Programs in ARS and in state experiment stations on

testing of new pesticides from industry should be
continued.
3. ARS research on basic insect biochemistry and
physiology and on mechanisms of resistance should be
increased.
4. Historically, ARS has had a relatively small effort
on weed control. This research and that on plant
growth regulation should be increased to a larger
fraction of the total pest control effort.
5. The present ARS technology licensing policy in effect
prohibits exclusive or semi-exclusive licenses of ARS
inventions. Nonexclusive licenses are satisfactory
in many areas of ARS research. However, they are not
attractive for pesticides for which large development
expenditures are necessary to obtain registration and
large amounts of capital required for a manufacturing
plant. The British National Research Development
Corporation is an excellent model for the licensing
flexibility that USDA needs. Specifically, both
commercialization of, and competition on, the important NRDC pyrethroids in the United States were
insured by license under NRDC patents to two and
only two companies.

NACA annually surveys its members on U.S. pesticide sales
and the related R&D expenditures (2). The members include
nearly all the companies with significant R&D on new
pesticides in the United States. In 1976, total sales in the
United States and export sales from U.S. production, were
$2.58 billion, up 5.1% from 1975, and the highest ever. Total
R&D expenditures were $195 million, up 25% from 1975. R&D
expenditures increased from 6.7% of sales in 1975 to 7.9% in
1976. Clearly, the industry is flourishing, is supporting a
large amount of R&D, and has the capability to find new pesticides. However, 11 of the 32 companies that reported both
R&D and sales had less than $10 million in sales. Average
R&D expenditures for these companies were 24% of sales. That
level of R&D can only be justified by the expectation of large
sales growth from new products. In addition, the number of
companies reporting R&D was down from 37 in 1975. These
surveys support other information that the smaller companies
are leaving the pesticide business or being acquired by larger
companies. Expenditures on synthesis and screening—the
innovative research on finding new pesticides—*decreased* from
33% of all R&D expenditures in 1971 to 19% in 1976. Actual
expenditures on synthesis and screening increased from $25
million in 1971 to $35 million in 1976, but the increase was
largely due to inflation. The most critical 1976 survey
figure related to R&D is that only four new pesticides

received first commercial registration in 1976. From 1967 through 1975, from 8 to 12 new pesticides were registered annually. In addition, the equally important registration of new uses for existing commercial pesticides has been greatly reduced. Skaptason of EPA recently reported (3) that only 50 registrations in total were approved in 1976 compared with 3,500 in 1973. The registration process is currently almost at a complete standstill.

The stalemate in EPA has created a discontinuity, a watershed beyond which the pesticide industry cannot plan with confidence. If a large number of important commercial pesticides on the potential RPAR list are not reregistered and market entry of new pesticides is delayed for several years, we must expect that only strong pesticide companies with large participation in markets outside the United States will continue R&D.

Clearly, the greatest, and indeed the only, needed incentive for U.S. pesticide companies to continue to provide the resources to discover and commercialize needed new pesticides is an effective, efficient regulatory process based on technically sound laws and regulations and implemented by a capable, technical staff knowledgeable in agriculture. By comparison, other incentives proposed earlier (4) are trivial. I will discuss only three of the main disincentives in the present regulatory system.

One is the reliance on arbitrary, formalized, legal and regulatory devices, such as RPAR and the continued addition of standard tests, which try to replace scientific judgment by long check lists tended by scientific clerks. These continue to increase R&D costs, tend to discourage good research and may encourage the registration of more toxic, rather than less toxic, pesticides. Consider two insecticides with equal insecticidal dosages and crop residues. Compound A, however, has acute oral toxicity to mice and rats of 100 mg/kg, while for B it is 10,000 mg/kg, or 100 times greater. The maximum tolerated dosages for lifetime feeding studies are 100 and 10,000 ppm in the food, respectively. Both compounds are potential weak tumorigens at 1000-5000 ppm in the food. The more toxic compound, A, cannot be fed at a high enough level to cause tumors and will be registered. The safer compound, B, however, will cause tumors in the trials and probably will not be registered, even though effect levels are perhaps 100,000 times greater than potential residues. These are hypothetical but realistic cases. There is common acceptance within the pesticide industry that compounds relatively safe in acute mammalian toxicity are more difficult and more expensive to register than compounds with relatively high acute toxicity. Acceptance by the pesticide industry of the Ames and other quick tests of mutagenicity has been delayed because of uncertainty as to whether a positive result in those tests can be overcome, in EPA's eyes, by clean long-term animal studies. Rather than a Delaney Clause, we need a national cancer assessment policy founded on the best scientific basis.

Second, neither EPA nor Congress has been able to cut the Gordian knot of public and private interests in the use of registration data by a second, "me-too", registrant. I agree with the three point NACA compromise proposal. First, following the example of human drugs, a detailed summary of this data on each pesticide, sufficient to satisfy the legitimate public concern but not sufficient for registration, would be made available by the company. Second, the first registrant would have a period of exclusivity on this expensive, manpower-devouring data of ten years, independent of patents. This is needed primarily for protection against use of the data by another company for registration outside the United States. Third, a me-too manufacturer should not rely solely on the first registrant's data but should demonstrate that his product is as safe as that first registered.

However, the greatest problem to the pesticide industry in registration of new pesticides is the delay and uncertainty in issuance of a label.

In an earlier, similar workshop, J.P. Rogers of Mobil Chemical Company pointed out the financial risks of delay in registration (4). Table 1 shows Rogers' financial assumptions for a hypothetical, *successful* new pesticide. Figure 1 shows the cumulative cash flow, as a solid line, for this pesticide. Rogers' curve has been modified to provide a first commercial label in year 5, after issuance of the patent, in agreement with industry experience. Let me quote

Table 1. Hypothetical Pesticide Cash Flow Assumptions

Price:	$3.33/lb
Volume:	10 million lb/year
Fixed Capital:	$20 million
Working Capital:	50% of sales
Sales General and Administrative Expense:	$3 million/year
Advertising and Other Expenses:	$1.5 million/year
Net Profit After Tax:	14% of sales
Cumulative (6 years) R&D Costs:	$14 million

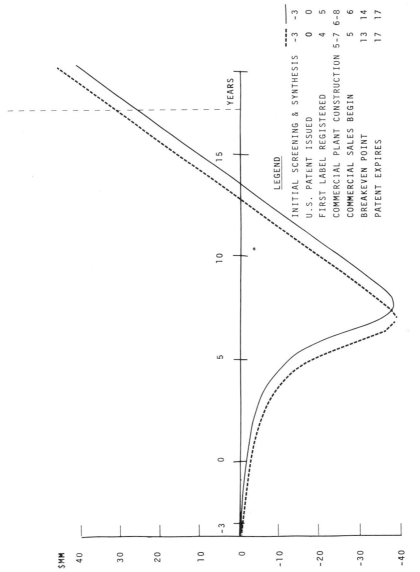

Fig. 1. Cumulative cash flow (million = MM) for hypothetical pesticide.

Rogers with appropriate changes:

> The point is that a company has some $40 million
> tied up in R&D and plant before significant commercial
> sales can begin. Then, if the product is successful—
> and that's still a risk because the grower decides the
> product's value, not industry and not government—if it
> is successful, then it's year 14 before you get your bait
> back and go into a positive cash flow position. By year
> 17, when the patent expires, if you're lucky, you have
> made a reasonable return on your money and effort. Of
> course, if the product is not accepted, the company has a
> disaster. A $40 million loss can seriously damage a big
> company. Such a possibility truly excludes small compa-
> nies and many large companies from participating in this
> area of business.
>   The risks have become too great for the rewards.
> Remember that our free enterprise system, which is the
> best in the world, demands a risk-reward balance. If
> resources can produce more at less risk in other areas
> of effort, that's where the resources go.

The broken curve shows the effect on cash flow of obtaining the first commercial label one year earlier. The break-
even point is now also reached one year earlier, at year 13,
and the positive cash flow before expiration of the patent is
increased by 25%. Proposals have been made to start the 17-
year clock of patent life for a regulated product at the time
of issuance of the first commercial label. This is of help
but more than four dollars income from extension of the
patent life are required for a present value of one dollar
from market entry a year earlier, and the risk is much greater. Additional delays of two or three years in obtaining
first registration clearly could have a disastrous effect on
the risk-reward ratio for new pesticides.
      I feel that pesticide registration, with the added
burden of reregistration, has become a nearly impossible task
because of the conflicting pressures from within Congress,
from environmental groups, and from industry. We are all part
of the problem but the largest factor is the basic character
of EPA.
      In 1970, responsibility for pesticide regulation was
transferred from USDA to EPA because USDA was biased toward
its constituency, the farmers, and did not adequately consider
environmental problems. Six years of experience has demonstrated that EPA is also biased and unsuitable. The reason
is fundamental. Pesticides are the only toxic chemicals
deliberately added to the environment. Allowing such use is

contrary to EPA's basic responsibility to remove pollutants, wastes, and chemicals from the environment and to the views of EPA's primary constituency, the environmentalists. Without responsibility for food production or experience with agriculture, EPA is, in addition, only able to assess the risks from pesticides and not to make the balanced risk/benefit assessment mandated by Congress. With these psychological conflicts in EPA, it would not be surprising if, as accused by pesticide industry critics, administrative devices are frequently used to delay or prevent registration rather than to register new pesticides.

If neither USDA nor EPA can provide the balanced, reasoned environment needed for pesticide regulation—where should it be placed?

I propose that registration of pesticides be the responsibility of a new Department of Food, similar to the proposed Department of Energy. This Department of Food is needed to provide overall responsibility for national food policy and for production and quality of the food supply. The constituency should be the consuming public—the total American public. I suggest an organization with respect to pesticides as shown in figure 2. The present Department of Agriculture would be part of the new Department of Food and would have responsibility for development of pest control techniques and for assessment of pesticide performance and benefits to the farmer and to the nation. The food half of FDA would become part of the Department of Food with responsibility for food quality. The Department of Food's registration organization, moved from EPA, would also be responsible for risk/benefit assessment. The safety and risk evaluation of pesticides would remain administratively part of

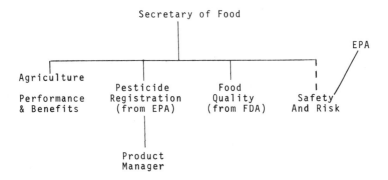

Fig. 2. Organization of proposed Department of Food.

EPA but with functional guidance from the Department of Food.
The total pesticide regulatory process would be linked by a
matrix system (5). Let's look at two levels. The product
manager of EPA should be transformed from a glorified clerk
to a true manager in the Department of Food's regulatory
organization with responsibility for registration (fig. 3).
He would function as head of a team of designated representatives from Agriculture, Food Quality, EPA, and Interior,
which would assess the total registration package of information and provide a balanced basis for decision, including
questions such as trade-offs between residues of insecticides
and insect parts in food. The team members report to the
product managers, functionally, but administratively to
Agriculture, Food Quality, EPA, and Interior, respectively.
At the top, the Secretary of Food would have final responsibility for registration and cancellation with advice and
guidance from Agriculture, Food Quality, EPA, and Interior.
This matrix system is used in industry to insure effective
working relationships between such functions as marketing,
manufacturing, and R&D under the responsibility of a product
or business manager. The nation needs a system such as this
to provide an effective, scientific regulatory process with
ultimate decision making by an official who is responsible
for the food supply and to the public as a whole. Hopefully,
this proposed Department of Food can provide the needed
balanced judgment on both pesticides and biological control
to optimize the supply and quality of food consistent with
environmental considerations.

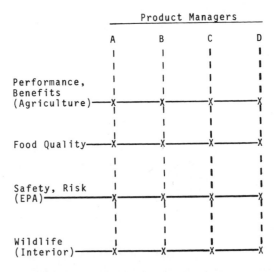

Fig. 3. Proposed organization to register pesticides.

REFERENCES

1. Bottrell, D.G. and P.L. Adkisson. 1977. Cotton insect
   pest management. Annu. Rev. Entomol. 22:451-81.
2. Ernst and Ernst. 1977. Industry Profile Study. Natl. Agr.
   Chem. Assoc. April. Mimeo.
3. Skaptason, J.L. 1977. Registration process is now
   essentially stopped. Pestic. Toxic Chem. News 5(19):14.
4. Rogers, J.P. 1977. Final Report. Proposed Incentives:
   Industry's Viewpoint. Arthur D. Little, Inc., NBS-GCR-
   ETIP-76-34. pp. 25-32.
5. Goggin, W.C. 1974. How the multidimensional structure
   works at Dow Corning. Harvard Business Review, pp. 54-65.

DISCUSSION

M. HINKLE: Would you comment on the fact that pesticide sales
have increased each year since 1945, despite increased registration costs, etc.; yet since 1976, several companies such
as Allied, have gone out of agricultural chemicals because
of their own activities. For example, the mirex registration
was sold to the state of Mississippi. Another company was
hauled in on Kepone, and has gone out of agricultural
chemicals.

R.R. WHETSTONE: Several companies attempted to become viable
in agricultural chemicals, but then stopped R&D; they
continued to sell existing products, but actually abandoned
pesticide business effectively about five years earlier.
Some small companies have been unable to come up with new
chemicals, and are on the way out. Government regulations
are only part of the reason; the highly competitive nature of
the business is also part of it. The increase in sales of
pesticides is due partly to inflation, to the replacement
of cheaper insecticides with more expensive ones, to the
development of new products, and to increased exports.

E. JANSSON: The competitivenes in the pesticide industry is
not so real, because most chemical companies are subsidiaries
of large oil companies. These companies have been charged
in court with not being competitive. Secondly, you charge
that there is a conflict of interest in EPA, but what about
the conflict of interest when pesticide salesmen give advice
to growers.

R.R. WHETSTONE: Nearly all pesticide companies are indeed
part of large chemical, drug, or oil companies, although they
operate as semi-independent businesses. I don't know what

your criterion is for competitiveness of a business, but there are 33 chemical companies that do research on pesticides, and none has more than about 10% or 15% of sales.

E. JANSSON: According to legal criteria, if the three largest firms have more than 20% of the sales, then the industry is suspect.

R.R. WHETSTONE: Then the pesticide industry is suspect, but it is intensely competitive in terms of R&D. As for "conflict of interest," USDA was rightly criticized for looking at pesticides from one point of view; I think EPA is just as biased, but with the opposite point of view. With respect to pesticide salesmen, the growers get advice from a number of different salesmen, just as you would if you were buying a car. Farmers also get advice from the experiment stations and other sources.

E. JANSSON: Isn't the choice between a Ford or a Chevrolet fundamentally different from the choice between pesticide or no pesticide?

R.R. WHETSTONE: It is a question of where the farmer gets his advice. Traditionally, the pesticide companies have worked with USDA and extension and the farmer, providing the tools needed by the farmer. I feel that in very few cases do we yet have proven reliable IPM methods to use. The farmer, in making his choice between pesticides and no pesticides has long experience and can get advice from government experts.

A. ASPELIN: With respect to problems with the registration process, the 1972 amendment was written in such a way that any agency would have had trouble implementing that mandate. It is not a problem inherent with EPA.

R.R. WHETSTONE: I agree.

R. CLARKE: Is it the position of Shell that the country will suffer serious losses of food and fiber without the use of dangerous chemicals?

R.R. WHETSTONE: My own position is that pesticide use in the last 20 years has contributed significantly to human health and our food supply. To back away completely from the use of pesticides without replacements for pest control would put many lives in jeopardy from decreased food supply.

T. CLARKE: But does Shell believe that you must allow the use of chemicals with demonstrated adverse effects on humans or their environment, for example, chemicals banned by the Delaney clause?

R.R. WHETSTONE: My point on the Delaney clause was that it is an arbitrary, legal attempt to do good; instead, I think we need a national cancer policy. I agree that we need to assess the risks and benefits, but we need to assess degrees of risk. Shell is concerned with health risks. For example, when EPA banned aldrin and dieldrin on crops, but allowed its use against termites, Shell withdrew the chemicals completely, for termites too.

E. CHOFFNES: Concerning the Delaney clause, is it your position that chemicals that cause cancer in laboratory animals are not necessarily carcinogenic in humans?

R.R. WHETSTONE: It is my personal feeling that animal studies are desirable, the best we have, and should be used. Where I differ is in their interpretation. Instead of a yes/no interpretation, I think we need to assess levels and types of tumors in making a decision.

E. CHOFFNES: In a review of 25 pesticides that have tolerance levels, 24 out of 25 were found to be carcinogens based on the industry's own tests. How do you interpret what is a safe level of cancer?

R.R. WHETSTONE: Cancer experts cannot answer that question— I certainly cannot.

E. CHOFFNES: Can the industry develop safer, more selective chemicals?

R.R. WHETSTONE: Selectivity in terms of safety to man and large animals vs. insects is clearly desirable, but the question of specificity to one insect vs. another is more difficult. We have developed specific pesticides, but we can't sell them right now. We need a different attitude on the part of farmers, who now want broad-spectrum pesticides.

UNIDENTIFIED: Are any of these specific pesticides registered?

R.R. WHETSTONE: Two of them are; one is now used on animals; one can be used against the corn rootworm, but is not being produced now because a better chemical has made ours noncompetitive. It will be difficult to interest industry in specific chemicals until there is an IPM framework for their use.

BARRIERS TO THE DIFFUSION OF IPM PROGRAMS IN COMMERCIAL AGRICULTURE

Wayne R. Z. Willey

Environmental Defense Fund
Berkeley, California

We are now well into a second decade of serious national and even international concern with the harmful effects of pesticides in the environment. The scope of this concern has broadened over the years to include not only the risks of carcinogenicity and mutagenicity in humans and toxic effects on wildlife species, but also threats to the stability of agricultural production through the energy requirements of chemical pest control and the depletion of genetic stocks among pest and beneficial species. Yet agricultural production continues to rely primarily on chemical pesticides to control yield- and marketability-threatening pests, including insects, pathogens, bacteria, fungi, nematodes, rodents, and weeds.

The use of pesticides in the U.S. environment has risen steadily over these same years, reaching an approximate 1.2 billion pounds by 1975 (1). Individual pesticides have been developed and subsequently phased out of use in pest control, but the total load continues to grow. Meanwhile, the overall demand for food will continue to increase as long as the population grows and economic development proceeds around the world. With a projected world population of 7 billion by 2000 (2), intensified efforts to bring new lands into production, and an ongoing loss of agricultural production of 35-40% from pests (3), the outlook for reduction of total pesticide use is not good.

Integrated pest management (IPM) has long been advocated as an alternative strategy to strict reliance on chemicals in pest control. That IPM represents an innovation in pest control technology relative to eradicative chemical control (4,5) is accepted widely. It is probably even safe to say that there is increasing agreement on a definition of IPM (6).

285

There is considerably less agreement concerning the conditions
under which IPM has been or will be successfully applied.
The following discussion focuses on the key factors that are
significant in the actual diffusion of IPM in commercial
agriculture. Some illustration with the experience in
California will, I hope, provide insight into important barriers to diffusion. In light of the threat of increases in
pesticide loads in the years to come, such barriers are
crucial factors in any attempts to mitigate the harmful effects of pesticide use in the United States.

PRECONDITIONS TO THE DIFFUSION OF IPM

    The application of IPM has four general sets of precon-
ditions. First, a specific crop/pest IPM technology must be
available. For example, specific technologies would include
pheromones, beneficial species, detection and sampling
methods, as well as narrow-spectrum pesticides. IPM technologies are relatively developed in such crops as cotton,
citrus, alfalfa, and deciduous fruits but remain unproven in
many other crops. While much valuable university research
has been and will continue to be performed on IPM technologies, discoveries by commercial practitioners are to be
expected at an increasing rate as diffusion proceeds.
Innovative diffusion of new technologies, as well as the
invention of new related technologies, has occurred at an
increasing rate as applications are made commercially in many
industries[1]. Thus, while it will always be important for
university research on IPM technologies to proceed, it is
also crucial that the diffusion process proceed as well so
that the chances of new discoveries in the field are increased. This becomes increasingly so as university
graduates with pest management training and an understanding
of scientific research are available to become involved in
commercial pest control.
    Second, IPM applications require scientific information
on pest and beneficial species, alternative controls, etc.
Data on a specific agroecosystem can only be obtained from
observation in the field. Such observation takes time and
knowledge, such as that involved in the monitoring of
Egyptian weevil in alfalfa or mites in citrus. This kind of
surveillance activity has become an important component in the
services performed by independent pest control advisors. Such

[1]*For industrial examples see (7); for agricultural examples see (8).*

advisors are discussed below.

Third, an IPM program requires that the technology and
information be distilled into decisions by an objective
practitioner. In the end, sound professional judgment is
crucial to successful application. This is true of most
commercial applications of scientific knowledge, particularly
in the life sciences. Objective means that the judgment of
the pest control practitioner—whether a grower, a farm
advisor, a chemical salesman, or an independent advisor—is
concerned solely with the optimum performance of the IPM
program. Such optimum performance is in the grower's
interest since it minimizes pest control costs while maintaining yields, i.e., the economist's notion of profit max-
imization. If the practitioner has other objectives that
conflict, then the IPM program's successful application is
jeopardized. Such objectivity, although more difficult to
measure, is nonetheless as important to IPM diffusion as the
availability of the technology or of sufficient information
in the field.

Finally, the application of IPM must not require unbear-
able economic risks to the adopter. Such risks involve the
costs and marketability of production. Although IPM involves
a reduction in pesticide costs, it often requires "factor
substitution"—the use of other inputs, such as labor,
instead of pesticides. On the production side, the maintenance of yields is required, particularly if there is no
significant net reduction in costs. Although the theory of
IPM indicates that, in most cases, a diversified use of inputs
to control pests will result in less reliance on pesticides,
it is really an economic and statistical problem to determine
the yield and cost consequences of IPM. The crucial economic
variables are (1) the relative costs of pesticides versus
other inputs and (2) the current prices of output. Obviously, the economic performance of IPM cannot be measured until
sufficient diffusion has occurred. In California, sample
survey data measuring performance of the IPM programs of
independent advisors[2] indicate generally that such programs
have not involved economic losses. In many cases, economic
profits from IPM application were recorded. A strong indication of the economic viability of IPM programs lies in the
increasing numbers of independent IPM advisors in California,
Arizona, and Texas, as well as in parts of the Southeast, the
Midwest, and Northwest. If such programs were not in the

[2]*For a 1970-71 cross section of California crops and
counties see (9); for a specific analysis of San Joaquin Valley orange and cotton programs in 1970-71, see (10).*

growers' economic interests, then presumably such growth
would not occur. The California survey (9;p.10) revealed the
proportion of acreage under independent advisors' programs to
be about 9% of total cotton, oranges, alfalfa hay, grapes,
and tomato acreage, with nearly 7% of total sugar beet, nuts,
deciduous fruits, vegetables, melons, and alfalfa seed acreage.

Further, assurance that the resulting production will be
marketable is necessary. For some commodities, such as
cotton and alfalfa, this is not a serious consideration. For
fruits and vegetables, however, market grading systems impose
serious constraints on the grower's ability to incur certain
kinds of pest damage. The problem of "cosmetic" uses of
pesticides—uses aimed solely at produce appearance and without nutrition or physical yield-inducing benefits—is
becoming a topic of increasing concern and controversy.
Recent estimates (3) indicate that 10-20% of all pesticide
use in fruits and vegetables is for cosmetic purposes.
Regardless of one's view of the definition or benefit of
"cosmetic" uses, it is certainly true that the adoption of
IPM in such crops is seriously impeded by certain quality
grades that accompany marketing orders.

Barriers to the diffusion of IPM presently exist in
commercial agriculture (as well as in urban pest control) because one or more of these preconditions are not met. In many
scientific circles, one often hears that we do not have
sufficiently developed IPM technologies and, therefore, more
basic research is needed. This was certainly a key conclusion
of the recent NAS study (11) on pest control. Although insufficient technology is often a serious problem, inadequacy
of information in the field, lack of objectivity of pest control practitioners or advisors, and the need to assure market
grade standards are also common problems. The barriers to
diffusion that presently exist in the areas of information/
advice and in produce marketability are discussed in the
following two sections.

TECHNICAL INFORMATION AND ADVICE IN THE FIELD

IPM programs can be produced either by private industry
or by government. Although various governmental programs have
been utilized (e.g., the USDA cotton scouting program via
various experiment stations and county farm advisors), the
long-term goal of such programs has been to let private
industry bear the burden. The traditional role of experiment
station and county farm advisor personnel in extending advice
and knowledge will likely continue. This role is supportive

to the efforts of growers and agribusiness. It influences
the behavior of agricultural practitioners and therefore
provides a service at public expense. Direct pest control
decision making, however, will likely remain the responsibility of private industry.

The major source of technical information and advice on
pest control in the private sector is the field personnel of
the pesticide companies. A minor source is the independent
(of pesticide companies) pest control advisor. Various
grower cooperatives are also sometimes a source of pest
control services. Some growers, particularly those with
large operations, retain their own pest advisors as employees
rather than consultants. The typical mode of operation for
pesticide company employees is to provide information and
advice to the grower, usually free of charge, and at the
same time perform pesticide sales. The independent advisor,
on the other hand, charges a fee for information and advice,
but does not sell a commodity. There are exceptions to both
of these rules, such as independent advisors also acting as
sellers of beneficial species.

Table 1 provides summary statistics (9;p.93) for 22 crop/
county combinations in California in which independent
advisors' programs are available. These statistics represent
the ratio, in each county and crop, of acreage under an independent advisors' program to acreage not under such a program.
Since the vast majority of licensed pest control advisors in
California are also employed by pesticide companies, it is
safe to say that most of the acreage not under an independent's program is serviced by a pesticide company employee.
The statistical mean over these five crops—cotton, alfalfa
hay, grapes, tomatoes, and oranges—is a ratio of 0.140, with

Table 1. Summary Statistics: Ratio of Acreage Under IPM
Programs of Independent Pest Control Advisors to Acreage Not
Under Such Programs. 22 Crop-County Combinations; California,
1972 (9;p.93)

Crop	Mean	Standard Deviation	High	Low
Cotton	0.185	0.151	0.450	0.042
Alfalfa Hay	0.110	0.062	0.165	0.022
Oranges	0.123	0.091	0.327	0.051
Grapes	0.040	0.038	0.121	0.003
Tomatoes	0.215	0.086	0.330	0.105
Overall	0.140	0.118	0.450	0.003

a high county/crop of 0.450 and a low of 0.003. This means
that an average of about 12% of all acreage in these crops
and counties was under an independent's program, with up to
31% in one case.

The growth in adoption of such independents' programs has
proceeded over a number of years. Figure 1 shows time series
trends of the ratio of independent to nonindependent acreages
fitted to the logistic growth curve. Curve I represents a
"high growth" case; Curve II represents a "no growth" case
(actually a declining share of the market); Curve III represents an average of 22 individual growth curves[3]. One conclusion is clear. Independent pest control advisory services
have been around in these areas and crops since 1957 on the
average, and as long as the early 1950s in some cases. In
other cases, the service has only been available for a few
years. The second obvious conclusion is that while the
"market share" of these independents has generally increased,
the rate of increase has been less than dramatic overall.
These conclusions are demonstrated further in figures 2a, 2b,
and 2c, which show growth curves statewide for an assortment
of crops. Again, while the overall trend is upward, it is
often at a slow and erratic rate. This lackluster performance is better understood by noting that of an approximate
$600 million agricultural pest control industry in California
in 1972 (around $800 million in 1977, an approximate 6% growth
rate), the independent pest advisors held a scant ½% of the
pie.

If independent advisors really do what they claim, i.e,
reduce pesticide costs more than the cost of their service
fee while maintaining yields, then the obvious query is why
this fledgling industry has failed to "take off" like other
cost-saving innovations such as computer services, tomato
harvesters, or, of late, home insulation. First, the advisors' service really involves a labor-for-capital substitution,
which is exactly opposite the prevailing trend of capital
intensification in agriculture and in developed economies in
general. Capital intensity requires, in most cases, cheap
and abundant energy. Increasing scarcity and cost of energy
is having some significant impacts on capital substitution.
These impacts will be increasingly felt in pesticide use.
However, energy scarcity has not become serious enough yet to
outweigh the historical advantages of capital-intensive pest
control in agriculture. Further, capital substitution is

[3]*This is an unweighted average. The 12% average figure
cited is weighted by acreage proportion of total for each
county/crop observation.*

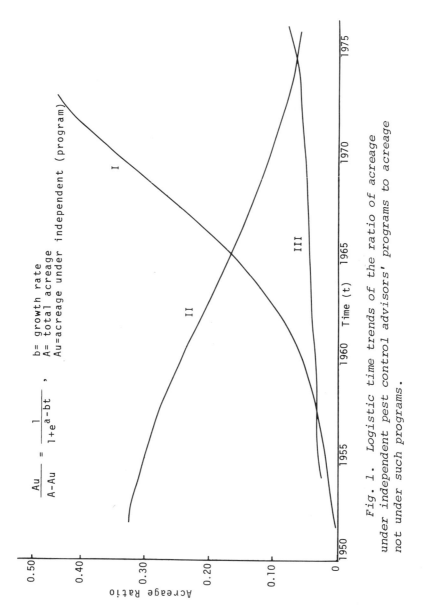

$$\frac{Au}{A-Au} = \frac{1}{1+e^{a-bt}},$$

b= growth rate
A= total acreage
Au=acreage under independent (program)

Fig. 1. Logistic time trends of the ratio of acreage under independent pest control advisors' programs to acreage not under such programs.

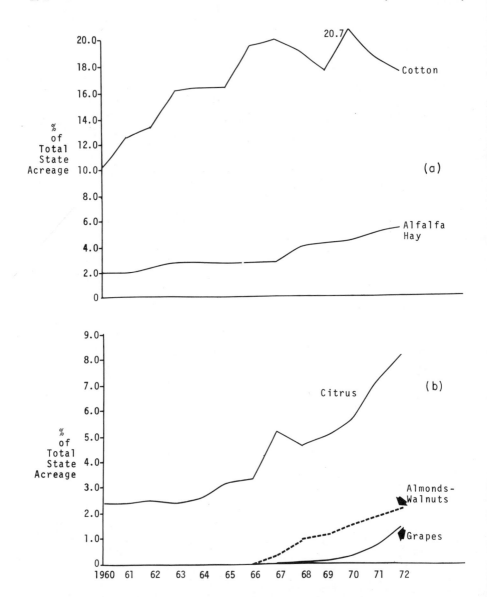

Fig. 2a. *Adoption of independent pest management consulting services—California: cotton, alfalfa hay (12).*

Fig. 2b. *Adoption of independent pest management consulting services—California: citrus, grapes, almonds-walnuts (12).*

Fig. 2c. *Adoption of independent pest management consulting services—California: alfalfa seed, sugar beets, tomatoes, vegetables, melons (12).*

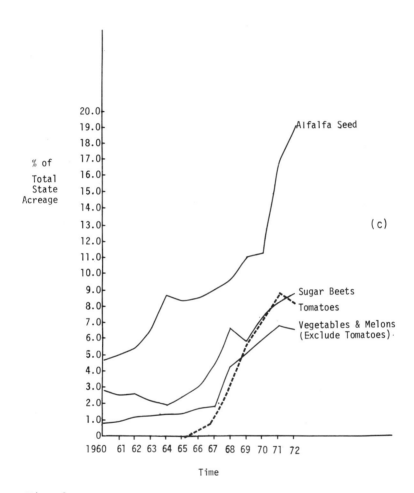

% of

Total
State
Acreage

20.0
19.0
18.0
17.0
16.0
15.0
14.0
13.0
12.0
11.0
10.0
9.0
8.0
7.0
6.0
5.0
4.0
3.0
2.0
1.0
0

Alfalfa Seed

(c)

Sugar Beets

Tomatoes

Vegetables & Melons
(Exclude Tomatoes)

1960 61 62 63 64 65 66 67 68 69 70 71 72

Time

Fig. 2c.

encouraged by rising labor costs. In agriculture, including
pest control, rising wages will likely continue to provide
incentive for capital intensity.

A second point, which is discussed further below, is the
fact that labor in general in developing economies has managed to regulate somewhat the process of capital intensification by unionizing in various ways. Union power has provided
a means by which labor could deal with uncertainties of the
market as well as the competition of capital. Even professionals like doctors and lawyers have their guild systems.
The independent pest control advisors face the worst of both
worlds—competition from a substitute (the pesticide industry)
and no organized power to deal with either management (the
grower, also relatively unorganized) or the competition (the
pesticide company, powerful and relatively organized).

Commercial Profitability of Independent IPM Advisors

Before exploring the power of the independents' competition in both the market and regulatory arenas, the economic
performance of the independents' programs deserves mention.
As was noted previously, perhaps the strongest indication of
the economic viability of the independents' programs lies in
the fact that they continue to exist and, on the whole, to
grow in a number of regions. Surveys in California over the
1971-75 period provide some of the most complete information
to date on the independents' programs.

Table 2 provides 1971 figures on prices of independents'

Table 2. Commercial Profitability of Independent IPM
Advisors' Programs in California, 1971 (12)

Crop	Revenue/Acre/Year ($)		Cost/Acre/Year ($)		Profit/Revenue Ratio
	Mean	Standard Deviation	Mean	Standard Deviation	
Cotton	4.69	2.14	2.34	1.63	0.50
Citrus	14.67	6.71	8.26	6.24	0.44
Alfalfa Hay	1.42	1.03	0.49	0.37	0.65
Grapes	3.38	0.96	3.32	1.86	0.02
Tomatoes	7.29	4.80	4.53	3.93	0.38
Sugar Beets	3.29	1.06	2.64	1.92	0.20
Deciduous/ Nuts	7.54	3.18	2.68	1.43	0.64
Alfalfa Seed	2.92	1.10	2.43	1.81	0.17
Vegetables/ Melons	5.68	4.65	1.56	1.02	0.73

programs (fee per acre per year) for a number of key crops
in California. These statistics are based on information obtained from 27 independent pest advisory firms in California
and represent the complete population of such firms at that
time (12). Included with prices of the services are per-acre
cost statistics. There is considerable variation in profitability both between firms and over crops. Nevertheless,
one sees that profits are the rule and that percentage of
revenue ranges from 2.0% to 73.0% with an unweighted average
statewide over all crops of 41.3%.

Since the independents' operations are highly labor-
intensive and therefore, are not easily measured in terms of
rate of return on equity, these profit-to-revenue figures are
the best available measure of rate of return. Profit-to-revenue is an indicator of economic incentives for entry and
expansion in a competitive industry. The existence of profits
in excess of managers' salaries under free entry and competitive conditions has historically resulted in increasing numbers and size of producers. This is the situation in the
independent pest advisory business, and one would expect that
growth in such businesses will continue to occur to the extent
that competitive conditions exist. The degree of competition,
of course, is determined in large part by the behavior of the
pesticide companies.

It is interesting to compare the profit/revenue ratios
for the small independent advisory companies to those of some
of their giant competitors. Table 3 lists the 1975 profit,
revenues, and resulting ratios of several corporations that
engage in pesticide production. The magnitude of the financial resources of oil and chemical conglomerates is striking.
The profit/revenue ratios are much lower than for the independent advisors. While this could mean a number of things,
it seems clear that the potential of and incentive for growth
is higher for the independent advisors on a dollar-for-dollar
basis. The size of the conglomerates, the relative fewness
of them, and the persistence of no entries of significant new
producers over a number of years have led to a history of
antitrust and divestiture proposals continuing through the
present. It is commonly recognized that the oil and chemical
industries are not competitive but there is little agreement
on what, if any, remedial policies are appropriate. While
this very complex subject is clearly beyond the present scope,
it holds relevance for an important issue: to what extent do
these conglomerates, through their agrichemical subsidiaries,
present a serious barrier to the diffusion of independent IPM
services? That is, even though the independent IPM advisors
look good in terms of earnings and could in theory be considered a "growth industry", they earned less than 1% of total

Table 3. Profitability and Sales of Major U.S. Corporations
Engaged in Pesticide Production, 1975 ($ Millions) (13-16)

Company	Profit	Revenues	Profit/Revenue Ratio
American Cyanamid Co.	148.0	1940.0	0.076
E.I. duPont De Nemours and Co., Inc.	272.0	7270.0	0.037
Exxon Corp.	2503.0	44865.0	0.056
Mobil Oil Corp.	810.0	20620.0	0.039
Standard Oil of Calif.	773.0	16822.0	0.046
Union Carbide Corp.	382.0	5665.0	0.067
United States Steel Corp.	560.0	8167.0	0.069
Allied Chemical Corp.	116.2	2333.1	0.050
Diamond Shamrock Corp.	114.2	1129.3	0.101
Eli Lilly & Co.	184.0	1233.7	0.149
Monsanto Co.	306.3	3624.7	0.085
Northwest Industries, Co. (Velsicol Chemical Co.)	101.1	1187.5	0.085
Stauffer Chemical Co.	98.7	949.8	0.104
Shell Oil Co.	514.8	8876.2	0.058

pest control revenues in California, where they are more common than in other states. Growth in the independents' operations must occur to some extent at the expense of pesticide
sales. Yet the institutional rules governing the pesticide
industry were established at a time when its growth was deemed
desirable by most segments of society. Some examples of these
rules in California are discussed below.

Commercial Performance of Independent IPM Advisors' Programs

    There have been a significant number of university re-
search projects that have demonstrated the potential of IPM
for reducing pesticide use under controlled field conditions.
However, there has been relatively little effort to date to
document IPM performance under commercial conditions. The
IPM programs produced by independent advisors offer the only
nonsubsidized market experiment in IPM application, aside from
various pest control cooperatives, which can clearly demonstrate the substitution of labor for pesticides. Some chemical companies have their version of IPM programs which, however, do not typically involve labor/consulting fees independent of pesticide sales. Some large farms hire their own IPM
advisors on salary, but there is little information on this
type of activity.

The 1971-75 survey of California's independent IPM advisors has provided some useful data. Table 4 provides summary statistics for random samples of San Joaquin Valley cotton growers in 1970 and 1971 who were either users or nonusers of independent advisors' services. The data indicate that for both growing seasons, the average yield was greater and the average insecticide cost less for those who employed independent advisors than for those who did not. When the insecticide costs of users are increased to include the independents' average fee per acre in this region [$2.68 (1970), $2.33 (1971)], the users' pest control cost is still less in both years. Table 5 presents similar results for San Joaquin Valley orange growers during 1970 and 1971. The same conclusion holds, even after adjusting for an average independent fee of $20.00 per acre: users' yields were higher and insect control costs less in both years.

These results are significant findings in that they support what one might have suspected by examining the post-1960 growth curves for cotton and citrus independent advisory services in California (see fig. 2). That is, since growers are certainly concerned more with the economic bottom-line of their pest control expenditures than with the environmental benefits of reduced pesticide use, the upward trend in the independents' share of the market implies that it is to the growers' economic interests to hire independents. Although the data and conclusions of Tables 4 and 5 need further expansion and refinement, they nevertheless appear to support the hypothesis that IPM programs are commercially viable in these crops in this region. From the independents' side, this hypothesis is supported by the profitability data of Table 2.

Expansion and refinement of the data in Tables 4 and 5 are presently proceeding in California (17). Larger statistical samples are being analyzed to increase estimate reliability. In addition, analysis of total farm management practices (i.e., fertilizers, irrigation, etc.) as well as environments (i.e., soil quality, climate) will allow independent statistical tests of the yield effects of pest control *vis-à-vis* other inputs. Again, this is all constructed in the context of samples of users and nonusers of independent pest advisors' services. Analysis of variance of yields over the 1970-74 period for cotton and oranges reveals year-to-year fluctuations in yield differentials of users and nonusers. Over the entire five-year period, one cannot reject the statistical hypothesis at a 95% level of confidence that the yields of cotton or of orange growers who hire independent advisors differ significantly from those who do not (18).

Studies such as the California surveys provide precise statistical information on the commercial viability of

Table 4. San Joaquin Valley Cotton: Average Yield Values and
Insecticide Costs Per Acre for Users and Nonusers of Independent Pest Control Advisors' Programs [see (10)]

Year/Category	Yield Value ($)		Insecticide Costs	
	Mean	Standard Deviation	Mean	Standard Deviation
Users, 1970	271.25	35.38	6.13	4.61
Nonusers, 1970	255.00	22.71	9.34	5.51
Users, 1971	281.93	20.51	4.21	4.72
Nonusers, 1971	221.65	72.93	11.97	7.38
Users, 1970-71	270.20	27.49	4.94	3.85
Nonusers, 1970-71	247.80	55.08	11.97	7.38

Table 5. San Joaquin Valley Citrus: Average Yield Values and
Insecticide Costs Per Acre for Users and Nonusers of Independent Pest Control Advisors' Programs [see (10)]

Year/Category	Yield Value ($)		Insecticide Costs	
	Mean	Standard Deviation	Mean	Standard Deviation
Users, 1970	527.17	237.22	24.58	13.89
Nonusers, 1970	509.47	187.36	45.64	19.42
Users, 1971	506.65	274.81	17.99	15.14
Nonusers, 1971	496.23	118.79	42.97	16.76
Users, 1970-71	515.80	260.64	20.53	14.97
Nonusers, 1970-71	502.85	157.00	42.35	18.29

IPM[4]. This information is of great use to policy makers in
Washington, D.C. and in the individual states, who will be
under increasing pressure, as pesticide loads grow, to reform
the field conditions under which IPM information and advice
are delivered.

MARKETABILITY AND QUALITY GRADE STANDARDS

For many food crops, particularly fruits and vegetables,
it is not enough that IPM can maintain physical yields while
reducing pest control costs. Even if the grower feels confident of this, he still needs assurance that the quality grades

[4]*For an investigation of IPM potentials in deciduous
fruits, see (19).*

that often accompany marketing orders from large buyers are met. The very essence of IPM is to manage, not eradicate, pest populations. This often results in the presence of *some* pests and *some* damage, which may violate quality standards. Such standards present a catastrophic risk to the grower, as the losses are nearly total and not incremental. That is, a fresh produce crop either complies with current standards on pest parts and/or damage or it does not. If it does not, the drop in value is dramatic. This is illustrated by the orange crop that fails to meet fresh market standards and is therefore sold for processing into juice or other by-products at around one-tenth of the fresh market value. This possibility presents an enormous risk compared to any yield reduction risks that might occur in, for example, the marketing of an IPM cotton crop.

Many high pesticide-using crops, including citrus and deciduous fruits, grapes, tomatoes, and other vegetable crops, will continue to present serious economic risks to growers adopting IPM as long as current quality grades and "cosmetic" pesticide uses are required. Various estimates of cosmetic use indicate from 10-20% of pesticide use in all fruits and vegetables (3) to over 50% for canning tomatoes (20). "Cosmetic damage" is commonly defined as "superficial damage to the exterior appearance of the commodity which damage does not significantly affect the taste, nutrition or storage capacity of the produce" (20). There is probably agreement on this general definition, but there is little agreement on the definition of specific terms such as "superficial" or "significantly," or even on the availability of IPM technologies in fruits and vegetables regardless of cosmetic damage (21). This disagreement will likely continue as long as there are difficulties in separating the cosmetic and yield damages of various pests, such as thrips in citrus or worms in tomatoes. These difficulties are exacerbated by the lack of experimentation with explicit controls on cosmetic pesticide use. There are clearly *some* cosmetic uses of pesticides, and such experiments could provide refinement in estimates of the extent of that use.

Even if statistical accuracy could be obtained in refining the 10% or over 50% cosmetic use figures mentioned, the consumer's behavior remains an issue. The cultural and economic factors involved here are complex indeed and many views have been expressed in untested hypotheses concerning this behavior. Untested hypotheses mean that regardless of individual views, there is little evidence of consumer behavior with respect to cosmetic damage. The problem is that the consumer has never had readily accessible alternative choices or information. The trade-off of pest damage or insect parts with pesticide residue hazards has never been available for the majority of U.S. consumers to make. That trade-off or choice has typically been prejudged by a combination of produce wholesalers, supermarket managers, and government regulators. For example, a consumer may be free to choose between competing brands of tomato paste with different prices, label colors, can size, etc., but he is not free to "shop around" for the "right" concentration of mold, worms, or pesticides. Although the tomato canner may reject the notion that consumers ought to have access to such unpleasant information, this is what is required to alter the present situation in which there is no readily accessible alternative to incurring unspecified risks of involuntary exposure.

The direct market behavior of U.S. consumers with respect to cosmetic use and damage will never be known without some market experimentation. Such experimentation is needed to provide evidence much in the same fashion that independent pest advisors' market experiments are providing increasing commercial evidence. As long as the present structure of food marketing is controlled by a combination of relatively few large wholesale buyers and quality grade standards that are focused on cosmetic damage and not pesticide use, such market experiments will be stifled.

A SCENARIO OF REGULATORY REFORM IN CALIFORNIA

It is clear that the diffusion of IPM in commercial agriculture faces a number of barriers, some of which are in the process of being removed and some of which are not. There is a growing body of evidence that individual IPM technologies exist and are economically viable. As mentioned earlier, this technical and economic feasibility is necessary but not sufficient for "take-off" of an IPM industry on a large scale in agriculture. The fact remains that even in the most successful cases of IPM adoption in California citrus and cotton areas, independent IPM programs have failed to capture a major share of the overall market.

A pesticide policy perspective begins with the hypothesis that IPM offers social benefits from reduced pesticide loads as well as technical and economic feasibility in certain pest control markets. The question then becomes one of specifying conditions that would allow widespread adoption of IPM. The following is a brief description of those conditions in the case of California. The underlying hypothesis is that the existing institutional structure of pest control in California is designed to facilitate chemical pest control and not IPM.

Marketing

The first major pest control decisions the grower makes in a season are the contractual agreements with a wholesaler or processor about quality grades. For many fruit and vegetable crops, this establishes the rules of the game for pest control until harvest. Although these rules are fairly specific for quality grades, there are no tolerance levels for cosmetic pesticide use. If cosmetic use were directly regulated, in as thorough a fashion as quality grading, the grower would feel constrained rather than encouraged to facilitate such use. There presently exists a variety of regulatory authorities to provide disincentives to cosmetic use. The following are some examples of heretofore unutilized authority in the State of California.

The State Health and Safety Code empowers the Department of Health to adopt regulations that prescribe tolerances including zero tolerance for poisonous or deleterious substances (H&S §26203). The Department must consider a full range of data on an individual substance, including its effects on animals and humans, and methods for its detection in food (H&S §26204). Although the State Department of Health has not historically been involved in pesticide regulations, there is certainly ample regulatory authority for it to do so with respect to pesticides in food. There is also presently political support for such involvement with pending legislation in the State Senate that would transfer explicitly much of the power to regulate pesticides from the Department of Food and Agriculture to the Department of Health.

With respect to consumer information on pesticides used or present in foods, the State Health and Safety Code again contains extensive language. The Code prohibits the dissemination of any false advertisement of any food (H&S §26460), the manufacture, sale, delivery for sale or offering for sale of any food that is falsely advertised (H&S §26461.5), and the receipt or delivery in commerce of any food that is falsely advertised (H&S §26462). Advertisement is false if "it is false or misleading in any particular" (H&S §26460). In determining whether or not an advertisement is misleading, all representations made or suggested by statement or word are to be taken into account, and the extent to which it fails to reveal facts concerning the food or consequences of its customary use are to be considered (H&S §26400). It would not be difficult to locate instances in which wholesalers or retailers are in violation here with respect to the pesticide aspect of advertised food.

The problem of food packaging and labeling is regulated by the Federal Fair Packaging and Labeling Act. Such

regulations are also the regulations of California, and the
State Department of Health may, when necessary, prescribe
packaging and labeling regulations for food (H&S §26438).
While innovations in processed food labels are possible with
respect to pest damage and pesticide use, no state authority
has been exerted here.

A final example of legal and administrative authority
that exists in California but has not been applied to the
problem of marketing of foods produced with pesticides, rests
in the State Penal Code. It is a misdemeanor for any person
to sell or offer for sale any food known to be adulterated or
which has become tainted or otherwise unwholesome or unfit to
be eaten with the intent to permit that food to be eaten (P.C.
§383). Adulterated applies here if any substance has been
mixed with the food so as to lower or injuriously affect its
quality or purity; or if by any means it is made to appear
better or of greater value than it really is; or if it contains any added substance or ingredient that is poisonous or
injurious to health.

These examples are taken from a limited part of the food
marketing-related regulations in one state. They illustrate
very well that the present practice of quality grade induced
pesticide use occurs not because of any lack of existing rules
but because of the power of the food wholesaling, processing,
and distributing industries which are able to play the game
by their own rules. This power persists in requiring the
grower to practice a chemically-oriented pest control, and
not IPM, in spite of the spirit and intent of the law and in
lieu of clear regulation based on that law. The existence of
this relatively concentrated market power (i.e., processing
and food chain corporations are relatively few in number) as
an intermediary between the relatively unorganized growers and
consumers presents an impediment to the free and competitive
exchange of food and information about food. It constitutes
a serious barrier to the diffusion of IPM.

Technical Information and Advice in the Field

The application of IPM techniques requires knowledge of
those techniques and current information on the crop and pest
ecosystem. That is, data and ability to interpret data are
required to produce an IPM program. In California, regulations related to both of these requirements exist. Informational requirements for pesticide use are contained in a
written "prescription" format. Qualifications for preparing
such "prescriptions" are governed by the "pest control advisors" system by which the state issues advisor licenses.

Reform of both information requirements and advisor qualifications are needed if IPM is to be encouraged. The present situation is one in which the information required to justify a pesticide prescription is inadequate and advisors licensed to make such prescriptions can have an economic bias in favor of pesticide use.

"Informational requirements" for IPM implementation include data on presence and densities of pest and beneficial species populations; stage in life cycles of these species; environmental factors (e.g., temperature, humidity) influencing birth/death rates of these species; stage of plant growth, including vulnerability and stress; alternative pest control measures available; etc. All pesticide use recommendations by licensed advisors must include a written "prescription" that requires certain information as listed in regulations promulgated by the California Department of Food and Agriculture.

The Food and Agriculture Code of California requires that the Director of the Department of Food and Agriculture enforce laws relating to pesticides and their usage. Divisions 6 and 7 of the Code broadly endorse IPM. Section 11501(f) of the Code makes it clear that it is state policy to support IPM "to achieve acceptable levels of control." The Director is charged with responsibilities regarding hazardous substances in several ways. Section 12824 orders the Director to ". . . endeavor to eliminate from use in the State any economic poison which endangers the agricultural or non-agricultural environment . . . ." Section 12825 provides that the Director may cancel registration of or refuse to register any economic poison that has uncontrollable adverse effects in the environment; that has less public value or greater environmental detriment than benefits from its use; or that has a reasonably effective alternative material that is less destructive to the environment. Section 14102 commands the director to "prohibit or regulate the use of environmentally harmful materials and to take whatever steps he deems necessary to protect the environment."

With this mandate, the Director is given power to cease operations and revoke licenses. Section 11708 allows the Director to suspend or revoke the license of any person who violates or permits the violation of Division 6 or 7 (Code provisions relating to harmful substances). Section 11737 (b) empowers the Director to order anyone to cease operations which are in violation of Division 6 or 7. With respect to pest control advisors, the Director's power to suspend or revoke licenses for violation of Division 6 or 7 is made explicit in §12023(d) of the Code.

To implement the pest control advisory system, Title 3

of the California Administrative Code requires of licensees
(1) applications and examination (§3120), (2) minimum qualifications (§3121), (3) registration (§3122), and (4) written
recommendations for applications of pesticides (§3123).

With regard to §3120-3122, California has begun an upgrading process to ensure some minimum competency in pest
control by examination and qualifications. The failure rate
of these examinations has been very low, and the qualifications targets in the early 1980s can be fulfilled in large
part by field experience in lieu of formal training. Although
this system does introduce some advisor standards, it is
difficult to argue that the quality of advice in the field
will be significantly affected given low failure rates and the
extensive "grandfathering" of the State's existing approximate
3,000 licensed advisors.

Perhaps of far more importance than the question of the
efficacy of examination and qualifications in promoting a
sound IPM advisory system, is the fact that the vast majority
of the state's present licensed advisors are also employees
of the pesticide companies. It is difficult to avoid the
hypothesis that IPM programs relying on less pesticides and
pesticide sales goals are contradictory. It is true that
IPM relies on some pesticides, particularly narrow-spectrum
types. Nevertheless, an approximate 6% growth rate in pesticide sales in the state is not an indication of the balanced
pest control of IPM, particularly given the very high usage
rates of materials such as toxaphene, carbaryl, methyl
parathion, and methyl bromide.

The failure to establish a program to eliminate conflict-
of-interest among licensed pest advisors presents a serious
obstacle to IPM. As long as the vast majority of advisors are
not objective in their judgments and assessments of risks, and
in fact have an incentive to bias the grower's perceptions
toward alleged "insurance sprays," effective IPM programs will
be difficult to achieve on a broad scale. While there is a
case to be made that pesticide salesmen are no more in con-
flict-of-interest than any other seller of merchandise, the
technical nature of the sales recommendation and the risks of
involuntary exposure incurred by third parties makes this case
tenuous. Recognition of the failure to institute a program for
phasing out the salesman/advisor conflict has resulted in
proposed legislation in the State Senate to prohibit this dual
role in California pest control advice (S.B. 669). Consequently, a long-overdue public record on the problem will be
established during 1978.

As for written recommendations (iv), §3123 requires that
each recommendation include such information as location of
property, crop/acreage to be treated, pest, pesticide name and

dosage rates, and reentry time information. Not even the very basic information need for IPM—pest density—is required in the "prescription." Obviously, these recommendation requirements are designed for chemically-oriented pest control and not for IPM.

For example, the control of codling moths in deciduous fruits can be achieved in part with IPM techniques. Pheromone traps for the moth adults are a useful means of monitoring the life cycle of the moth so that chemical applications, if any, can be timed during the reproductive phase. Density/age distribution data are needed for this but are not required in the recommendation prescription. Further, the potential of biological control by parasitic species has been noted. No information on such parasites is required on the recommendation. Finally, yield reductions from the moth occur at varying rates during the fruit development cycle, i.e., higher densities can be sustained with no real loss in yield during certain portions of the growing season. No information on fruit development/orchard conditions is required in the advisor's recommendation.

Thus, the provision of technical information and advice in pest control is governed by institutional rules that facilitate chemical, not integrated, control. Where concentrated industries, in this case oil/pesticides, control the flow of materials as well as information to a relatively unorganized buyer (economists call this "oligopsony"), this result is predictable. It consequently presents a serious barrier to the diffusion of IPM.

CONCLUDING REMARKS

The diffusion of IPM in commercial agriculture is becoming increasingly feasible both technically and economically, but continues to be hampered by regulations that support chemical pest control. The public policy debate on IPM will increasingly shift to this regulatory arena as evidence on feasibility becomes more diverse and refined. Eventually, we will have to confront the imbalances in economic power controlling the information and choices available to both the growers and the consumers. As long as the sales of the pest control, food processing, and food distribution industries can continue to grow at a reasonable rate and corporate equity is held intact, there is little reason to expect major changes in the direction of this momentum. Certainly, some companies will enter and leave these industries, but as long as earnings are reasonable, pesticide and food industries will likely continue to behave as they have.

A capitalist economy is based on the incentive to compete for earnings, and competition creates favorable conditions for new innovative ideas to surface and diffuse. The problem in pest control and food processing and marketing, however, is that competitive conditions are hard to find. A substantial restructuring and enforcement of the rules by which these industries are regulated is imperative if the economic and environmental benefits of IPM are to be realized.

REFERENCES

1. United States Department of Agriculture. 1975. Agricultural Statistics 1975. U.S. Govt. Print. Off., Washington, D.C.
2. Food and Agriculture Organization. 1974. Assessment of the World Food Situation. World Food Conference, Rome.
3. Pimentel, D., E.C. Terhune, W. Dritschilo, D. Gallahan, N. Kinner, D. Nafus, R. Peterson, N. Zareh, J. Misiti, O. Haber-Schaim. 1977. Pesticides, insects in foods, and cosmetic standards. BioScience 27:178-85.
4. Huffaker, C.B. 1971. Biological control and a remodeled pest control technology. Tech. Rev. 73(8):30-37.
5. DeBach, P. and M. Rose. 1977. Upsets caused by chemical eradication. Citrograph 62(6): April.
6. Smith, R.F. and R. van den Bosch. 1967. Integrated control. pp. 295-340 *in* Pest Control: Biological, Physical, and Selected Chemical Methods. W.W. Kilgore and R.L. Doutt, eds. Academic Press, New York.
7. Mansfield, E. 1968. Industrial Research and Technological Innovation—An Econometric Analysis. W.W. Norton, New York.
8. Tolley, G. (ed.) 1970. Study of U.S. agricultural adjustments. Agr. Policy Inst. Ser. #48, N.C. State Univ., Raleigh.
9. Willey, W. 1974. The diffusion of pest management information technology. Chapter 3. Ph.D. thesis, University of California, Berkeley.
10. Hall, D.C., R.B. Norgaard, P.K. True. 1975. The performance of independent pest management consultants. Calif. Agr. 29(10):12-14.
11. National Academy of Sciences. 1975. Pest Control: An Assessment of Present and Alternative Technologies. Vol. I. Contemporary Pest Control Practices and Prospects: the Report of the Executive Committee. Washington, D.C.
12. Willey, W. 1974. Pest management consultants in California —a case study. Presentation to VIII Annual Meeting, Assoc. Appl. Insect Ecol.
13. Standard & Poor's Corporation. 1975. Standard and Poor's Stock Reports—New York Stock Exchange. New York.

14. Anonymous. 1976. Annual directory of American business. Forbes 117(10):75-138.
15. Anonymous. 1976. The Fortune directory of the fifty largest commercial-banking companies, life-insurance companies, diversified-financial companies, retailing companies, transportation companies and utilities. Fortune XCIV(1):202-17.
16. Anonymous. 1976. The Fortune directory of the 500 largest industrial corporations outside the U.S. Fortune XCIV(2): 231-42.
17. Hall, D.C. In ms. An evaluation of alternative pest management implementation strategies. Draft final report, for Environmental Protection Agency.
18. Hall, D.C. 1977. Feasibility and implementation of integrated pest management in California agriculture. Presentation to XI Annual Meeting, Assoc. Appl. Insect Ecol.
19. King, N.B. and A.D. O'Rourke. 1977. The economic impact of changes in pesticide use in Yakima Valley orchards. Washington State University, Coll. Agr. Bull. 841.
20. van den Bosch, R., M. Brawn, R. Garcia, C. Magawan, A. Miller, M. Moran, D. Pelzer, J. Swartz. 1976. Investigation of the effects of food standards on pesticide use. Draft report, Environmental Protection Agency.
21. Council for Agricultural Science and Technology. 1976. Review of "Investigation of the effects of food standards on pesticide use." CAST Rep. No. 55.

DISCUSSION

W. LOCKERETZ: The means you gave for yield and insecticide costs of users vs. nonusers of independent pest control advisors are nowhere near being statistically significant. You should give the probability level for significance.

Z. WILLEY: The data is statistically significant even though there is a small sample size. The samples are small because this type of data is expensive to collect. We can calculate t statistics from the presented data.

M. EDEY: If these advisory firms are successful, why haven't they expanded? There must be obstacles—what are they?

Z. WILLEY: There is not free competition. The very large chemical companies dominate the information process in the field, and wholesalers set "quality" standards. The growers can't combat that because they have no choice of where to sell their produce. The consumers also have no choice concerning cosmetic standards. These are set mostly by the processors.

W. OKLOWSKI: The growth of independent advisors may be slow

because it is more difficult to sell a service than a product.

Z. WILLEY: Yes, and the chemical companies generally don't charge for their "service;" their advice is free. This causes problems for independent advisors.

T. CLARKE: Has the change in the price of pesticides changed the willingness of growers to use advisors?

Z. WILLEY: My own guess is that pest control cost is usually such a small portion of the total farm budget that the increases so far have not been enough to make the growers really feel a squeeze.

INTEGRATED PEST MANAGEMENT NEEDS—TEACHING, RESEARCH,
AND EXTENSION

Edward H. Smith

Department of Entomology
Cornell University
Ithaca, New York

It has been a challenging experience to listen for two
days to the varied points of view expressed, knowing that at
the end you were expected to glean, to synthesize, and to
offer some wise counsel on needs and directions for the future.
At this point I feel that the banquet to which Dr. O'Brien
likened this conference has gone on too long. I am stuffed
and suffering from indigestion. Despite this discomfort from
IPM overindulgence I will seek to offer you a demitasse—half
a cup, as contrasted to O'Brien's hors d'oevures.

Several noteworthy points have been established in the
course of this conference. Clearly, a group of leaders representing diverse views and backgrounds can come together in a
university setting and engage in a fruitful exchange of ideas.
As elementary as this step may be, it is a prerequisite to
progress and has been a long time in coming.

No one has proposed the banning of all pesticides. All
have recognized the "mixed blessing" characteristic of pesticides and the complexity in assessing short-range versus long-
range gains in pest control. There is unanimity regarding
integrated pest management (IPM) as the soundest strategy for
pest control. At least among ourselves, there appears to be
some progress in the understanding component of our theme.
The action component still lies ahead.

It is essential to consider the needs of IPM in the
larger context of the social, economic, and biological parameters impinging upon it. Previous speakers have established
these, but a brief summary will be provided here to set the
stage for our discussion.

THE HARD REALITIES

By now, informed individuals realize that pest control is
not merely the concern of those who happen to be in the business of farming. Rather, it is a matter of the utmost importance to the general public, a matter which strikes at such
critical issues as food to avert starvation, the health of
millions of people, and the viability of the environment on a
global scale. We can ill-afford big mistakes, and it is
extremely important that we properly assess the hard realities
that bear upon the problem.

The most critical issues are: population pressures, the
world food supply, energy, and viability of the environment.
The population curve, which cannot be quickly altered except
by catastrophe, indicates a world population of 6-7 billion
by the year 2000. The food requirements for this population
must be met largely by greater production on land now under
cultivation. The production technology that has accounted for
our impressive gains of the past few years is faltering.
There is concern that the fruits of technology may carry a
high price in indirect costs. In addition, current pest control technology requires high inputs of energy and poses
threats to the environment. These concerns come at a time
when the U.S. public has other concerns: rise in cancer rate,
the saccharin debate, Kepone pollution of the James River,
confidence in government, crime—concerns that feed upon each
other and lead to a vague feeling of an eroding quality of
life. In a free society these anxieties find public expression, and by the political process lead to the establishment
of institutions charged with responsibility to safeguard the
public interest. EPA was born of such concern.

There is a sense of urgency. The debate has extended for
a long time. IPM has emerged as a concept consistent with
sound ecological principles. The need now is for understanding and action, the theme of this symposium.

PARTICIPANTS IN THE DRAMA

Although the matter of pest control affects all the
people, key roles devolve to certain groups. Each of these
has built in biases, credibility problems, and different moti-
vations. Unless they can engender a new spirit of reexami-
nation and compromise, the outlook for progress is exceedingly
dim. There are signs that the participants in this extended
debate are drawing closer in support of common goals. The
characteristics of participants in the drama are evident.

The Farmer

Today's farmer is essentially a businessman, a manager of
resources. His survival depends on his ability to compete.
He is not necessarily a good ecologist; he is usually unwill-
ing to trust his own judgment in technical matters of pest
control and therefore seeks outside advice.

Agencies of Government

The two primary agencies of government with responsibilities in pest control and pesticide regulation are EPA and
USDA. Their respective roles set the stage for conflict. EPA
is a young agency (since 1970) granted tremendous authority,
operating under general statutes that are ambiguous and untested. The regulatory and enforcement mentality seems to
have predominated in the agency's early years and some public
backlash has occurred in response to all the major programs:
clean air, pure water, etc. Widespread criticism of EPA has
prompted a number of studies of its operations, the most comprehensive undertaken in 1973 by the Commission on Natural Resources of the National Academy of Sciences, on the recommendation of the House Appropriations Subcommitte on Agriculture,
Environment and Consumer Protection. The first in a series of
reports has been made available: "Perspectives on Technical
Information for Environmental Protection" (1). The report on
Pesticide Decision Making is expected to be released in the
near future.
The USDA, an old agency, is suspected of taking an extreme advocacy role for farmers. The quality of its research
and its interaction in the scientific community also came
under fire in an earlier National Academy study designated the
Pound report (2).

The Universities

The land-grant universities bear the major responsibility
for teaching, research, and extension in the area of pest control. These institutions have lacked credibility with environmentalists because of their role in developing and advocat-
ing chemical pest control. Their affiliation with growers and
the pest control industry is viewed by critics as too cozy to
be healthy.

The Chemical Industry

Its critics feel that the profit motive, which must be
industry's guiding light, is not compatible with judicious use
of pesticides. This distrust is fired by continued examples
of irresponsible action, which occur despite the regulations
designed to prevent such injury to people and the environment
(Kepone, occupational injury; Mirex pollution of Lake Ontario;
PBBs, dairy cattle in Michigan).

Special Interest Groups

The great surge of ecological awareness has produced many
special interest groups whose interest and motivation conflicts with those of agricultural producers. These groups
generally have no direct economic stake in pest control and
find little common ground with the agricultural producers.
These brief comments are sufficient to establish that
with respect to pest control we have a major problem with far-
reaching consequences, beset by conflicting views, institutional rivalry, insufficient knowledge, economic stress, and
a sense of urgency. Mr. Grumbly has discussed a number of
these issues, and his thought-provoking analysis deserves
careful study (see p. 261).
With the foregoing briefing to set the stage, let us now
focus our attention on the needs. I shall consider these
largely in the context of the land-grant university and its
programs of teaching, research, and extension.
At the outset we should acknowledge the unique role of
the university in the affairs of a free society. Its role is
to sense need, acquire knowledge to meet the need, and point
the way in applying the knowledge to the solution of the problems. The university is at its best when the three functions, teaching, research, and extension are interacting with
each other and providing catalytic feedback. It is this relationship—only vaguely understood by many—that represents the
uniqueness of the land-grant university. These functions are
not operating in harmony in advancing IPM as I shall indicate.

TEACHING NEEDS

In viewing teaching as it applies to pest control let us
recognize a responsibility both to liberal education and to
specialization in pest control. The former seems to have received far less attention than it deserves.
The problems of pest control will be with us for a long

time, and they will involve matters of public policy. The
development of sound pesticide policy will be virtually impossible unless the public can be won to some ecological middle
ground between the extremist views of some environmentalists
on one hand and the equally extreme views of some pesticide
advocates on the other.

If we consider the goals of liberal education, we can
hardly consider a person educated who does not comprehend, in
a general way, the biological processes regulating the flow
of energy through ecosystems, population dynamics, and ecological principles relating to environmental quality. These
topics, while they sound complex, are readily comprehensible
by students meeting the standards of good high schools and
colleges. Such understanding is clearly inherent in the concept of liberal education, at least as held by former President Perkins (3) of Cornell University. He stated that ". . .
the threefold purpose of liberal education is to learn nature,
society and ourselves; to acquire certain skills such as clear
expression and a grasp of the scientific method and discipline; and finally to embrace certain values, such as intellectual honesty, tolerance and the capacity for sound judgment."

It is clearly evident that college students are receptive
to training compatible with this concept of liberal education.
A wealth of teaching material is available and meeting the
need poses no unique problems. At Cornell a course in biology
for nonbiology majors, taught by one of the faculty's most
distinguished teachers, is being enthusiastically received.
Many opportunities exist for informal education along these
lines through the various programs provided for continuing
education—"life-long learning." The current wave of interest
in home gardening, house plants, and "green-thumb" activities,
provides the "teachable opportunity" for such training.

There are few biological models better adapted to the
teaching of ecological principles than insects, but these
appealing possibilities have not been well used by those involved in curriculum development.

Opportunities for deliberations toward moderate views
of the interaction of humans and living organisms and the need
for pesticides are constantly arising. For instance, the
occurrence of Rocky Mountain spotted fever on Long Island, an
area noted for environmental activists, has focused attention
on the relationship between humans, dogs, ticks, and their
natural wild hosts. Similarly, outbreaks of encephalitis have
required value judgments regarding human health, pesticides,
and environmental impact.

The point needs to be made that educational institutions
have a responsibility to liberal education, and that this

responsibility cannot be met without including teaching that
would contribute to "the capacity for sound judgment" in
environmental matters, including the issue of pesticide use.
Educational institutions caught up in the passion for specialization have not done justice to this part of their responsibility.

As we look ahead to the growing pressures for increased
agricultural production, the slow progress in finding alternatives to chemical control, the rising expectation of freedom
from insect annoyance and disease transmission, it is clear
that the development of public understanding and adoption of
manageable regulations is a major challenge.

Retread Training

There is an urgent need to reorient the outlook of individuals who received their training before the concepts of IPM
emerged. The concepts of pest management represent a radical
departure from the narrow disciplinary training that was the
norm for individuals who earned degrees in the 1940s and
1950s. It is individuals who received their training in this
period who are now in positions of administrative decision
making. Our system of "self-renewal" has not provided well
for updating the concepts of administrators, investigators,
teachers, and practitioners. There is an acute need for nondegree training programs for these groups. This need has been
recognized to a limited extent, and at least one session has
been directed at updating the views of administrators (4).
The groups that might be considered for nondegree refresher
training have been identified by the Glass report (5) as
administrators, researchers, teachers and extension personnel,
field survey personnel, pest control applicators, consultants,
and growers. I would emphasize the need for such training for
the field staffs of the state departments of agriculture and
Animal and Plant Health Inspection Service (APHIS), USDA. The
combined staff of these agencies represents a nucleus of
horticultural inspectors and others who can play a significant
role in detection and control of pests. A radically different
approach will be needed to provide meaningful training for
these and similar groups. Unless a decision is made at top
administrative levels to invest in time and funds for training, no appreciable change will occur. Mechanisms for coordination and implementation exist in the State Pest Management
Committees established by agreement with the Administrator,
Cooperative Extension Service. Another coordinating group is
the Advisory Committee to the Survey Entomologist. Such a
position has been established in most states, partially funded

by APHIS. In the cases cited the emphasis is primarily on entomology rather than the full spectrum of disciplines that should be considered.

Degree Training

Progress in teaching IPM has lagged behind progress in research and extension. A number of factors account for this. There is an expected lag time between discovery of new knowledge (research), application to practical ends (extension), and incorporation into the conventional wisdom and transmission to students through the teaching process.
A contributing factor is the lack of agreement regarding the concepts. Entomologists still argue over whether IPM is new or merely old practices with a new name. (It's some of both!) The confusion spreads to other disciplines being asked to join the integrated effort. The status of IPM is indicated in part by the fact that only one volume designed as a text has appeared to date. This volume, *Introduction to Insect Pest Management* by Metcalf and Luckmann was published in 1975 (6). This volume specifically concerns insect pest management; a volume to integrate the required disciplines is yet to come.
Another problem in teaching IPM is that in many departments teaching responsibility is assigned to a few individuals whose heavy teaching loads preclude their involvement in research and interaction with new developments in IPM.
There has been a lack of innovation in teaching IPM due in part to lack of support. Effective teaching of IPM requires work with live populations in developing life tables and injury thresholds, use of computer technology, and field experience. To add these features increases cost of instruction. Another problem in some localities is the academic calendar, which limits opportunity for field work before frost in the fall and after the spring thaw.
Today's students have special needs not common in the previous generation. Many are from urban backgrounds and have had little practical experience in the care of plants and animals. Innovations in providing intern experience is essential.

B.S. Degree

Training is needed for the student who seeks no training beyond the B.S. and for the student whose goals include the M.S. or Ph.D. It appears that these two educational needs are

not well met in the same institution. The emphasis is either
on graduate training and the research philosophy or on applied
training. One is usually secondary to the other.

The B.S. programs in plant protection that attempt to
provide the interdisciplinary training required for careers in
practical pest control have not met with general success.
Invariably the disciplinary bias of the faculty advisers has
prevailed, and the students have emerged with essentially
an undergraduate major in entomology or plant pathology.
Despite the urgent need, there has been relatively little
support for training grants. An innovative program is currently under way at Michigan State University with NSF sponsorship.

The needs of the terminal B.S. student and the predoctoral
student are quite different, and the latter may be best served
by a biological science major with specialization at the
graduate level.

Professional Master's Degree

The need for a professional master's degree for prac-
tioners of IPM has been recognized by a number of institutions. Simon Frasier University in British Columbia has
established a Pestology Center, which offers a degree of
Master of Pest Management. Requirements for the degree include the writing of a professional paper, field experience,
and courses selected from three groups designated "Organisms,"
"Controls," and "Processes." A useful description of the
program has been provided (7).

Ph.D. Degree

There are two views regarding the orientation of training
for the Ph.D. One holds that the core of training should be
in one of the long-recognized disciplines with minors in other
disciplines to provide the needed interdisciplinary perspective. The other view holds that a new kind of technical
specialist is required whose primary orientation is not to a
single discipline but to the array of disciplines that impinge
on the many facets encompassed by pest management.

Opposing views regarding the Ph.D. degree are represented
in the Glass report (5) and the "Workshop on the Professional
Doctorate in Pest Management" (8). The former states that
"individuals would receive their degrees in a specific crop
protection field (entomology, plant pathology, nematology, or
weed science) with minors in at least two other fields. Their
major area within that field would be pest management." By
contrast, the advocates of a professional doctorate in pest

management propose ". . . a broad background in crop production, economic entomology, weed science, plant pathology,
vertebrate zoology, ecology, pesticide chemistry, environmental law, and economics. To be able to deal effectively
with the public, he must also have an appreciation of human
relations, sociology, psychology, and communications." The
report goes on to state that "at no point will the level of
training for such a professional approach that of the disciplinary specialist in any particular field." And further:
"It is unlikely that this new kind of specialist can be educated within the existing programs of most universities." A
further elaboration supporting the professional Doctorate of
Pest Management is provided by Peterson (9).

I do not agree with the proponents of the Doctorate of
Pest Management, at least not at this stage of development.
The scope of subject matter to be encompassed by the pest
management doctorate (listed above) is far too broad for anything but superficial coverage. There is no substitute for
specialization, and our chances of adding the necessary interdisciplinary perspective to well-trained individuals in the
various disciplines who then join in team effort is far better than looking to generalists who are limited in compre-
hending the place of any one discipline in the total effort.

The outlook for a new breed of well-trained representatives of disciplines who are oriented to team effort is far
better than is generally recognized. A decade or so will
remove by retirement a layer of senior workers whose mind-set
was shaped before the current concepts emerged. One gets
glimpses from younger faculty members and particularly from
graduate students of the possibilities for exciting partner-
ship without the affliction of the traditional mental and
physical compartmentalization by disciplines.

It is beyond the scope of the present assignment to consider the details of curricula content for degrees in IPM.
Useful background information will be found in reports by
Browning (10), Glass (5), Obrien and Matooka (11), Matooka
(8) and Ventre (12).

Factors that have a negative influence on curriculum
planning and student specialization are the uncertainty regarding the number of individuals needed and who will be their
employers. This matter was considered by the Kennedy Task
Force (13), which offered the recommendation that "the USDA
and Cooperative Extension Service (CES), . . ., develop
estimates of the need for future pest management services and
programs for training the personnel necessary to implement
them." This is an exceedingly difficult assignment, considering the transition in both scope and sponsorship of IPM programs.

RESEARCH NEEDS

National Perspective

Research in IPM is conducted in both the private and
public sector. In the era of chemical control the teamwork
between the public and private sector was very effective in
meeting the needs of the pest control enterprise. Expressed
in its simplest terms, pest control relied chiefly on chemical
pesticides and industry met the need. The needs of IPM cannot
be met by this simple formula; therefore, a reassessment of
roles and responsibilities is in order in both the public and
private sector.

A fundamental question is whether funding through the
public sector is realistic in terms of need. It is useful to
compare the level of funding for agricultural research with
other research programs in the national interest. The following is adapted from the Kennedy report (13;p.157).

Resources Allocated to Agriculture and Pest Control
Research as Compared to Total National Research Resources
(1970).

Research Area	Funds as % of National Total
Defense	30.2
Space	13.5
Chemical Industry	6.1
Agriculture	2.0
Pest Control (including pesticide industry)	0.7

These figures suggest that the support for agricultural
research is relatively modest as compared with support for
some other national goals. The relativity of these figures
raises pertinent questions regarding motivation in establishing national goals and providing support. In light of the
hard realities cited earlier, these levels of support seem
disproportionately small.

In the political process, the advocates for agricultural
research support are largely the representatives of agricultural states. The important stake of the general public in
the agricultural enterprise has not won general political
support due in part to impressions carried over from periods
of excess agricultural commodities.

The funds in support of agricultural research in the
public sector in FY 1972 were divided 42% USDA and 58% state
agricultural experiment stations. Of the total, 21%, amounting

to $132,000,000, was devoted to new pest control technologies
(13;p.151). As will be discussed later, administrative constraints pose some problems in utilization of these funds.

Guide to Future Programs and Needs

The matter of pest control has been in the public eye for
a number of years and several comprehensive task force reports
have been provided as a guide to planning. These include the
following:

1963   The Use of Pesticides, a report by the President's Science Advisory Committee, Washington,
D.C.

1965   Restoring the Quality of Our Environment. Re-
port of the Environmental Pollution Panel,
Presidents' Science Advisory Committee,
Washington, D.C.

1966   Scientific Aspects of Pest Control. National
Academy of Sciences Publication 1402.

1969   Principles of Plant and Animal Pest Control.
National Academy of Sciences Publication 1965.

1972   Pest Control Strategies for the Future. National Academy of Sciences.

1975   Contemporary Pest Control Practices and Prospects. Vol. 1, National Academy of Sciences.

These reports have in general been useful in identifying
the deficiencies of past programs and in pointing the way for
corrective action. The latter source provides a particularly
comprehensive and helpful perspective. Another pertinent
though less comprehensive report is the special publication of
the Entomological Society of America, "Integrated Pest Management: Rationale, Potential, Needs and Implementation" (5).
This report was provided through support made available by the
Office of Pesticide Programs, EPA. From these reports and
more recent assessments, some conclusions can be drawn regarding prerequisites for research to advance IPM. Some of
these are:

1. Research goals in IPM must be reconciled with social,
economic, and biological factors. The economic factors are
obvious enough. The private sector must operate in
accord with the profit motive and the public sector is limited
by the funding available. Several far-reaching biological
factors require consideration. Insect resistance, for instance, is merely an expression of the principle of natural
selection, the selective agent being a toxic synthetic compound. The concept of the niche, species displacement, etc.,
all operate as ecological principles in the man-made

agroecosystem.

The social considerations will be paramount in the future. As we have already seen, certain pesticides are unacceptable because of their environmental impact or hazard to human safety.

2. The present knowledge base is inadequate for the imlementation of comprehensive programs of IPM.

The knowledge base from which IPM programs should be projected is woefully inadequate. We cannot cite a single commodity for which our knowledge permits full exploitation of the concept of IPM. Even with the crops that have received most attention, such as cotton, we are still a long way from the needed knowledge base.

When one considers the limitations in our knowledge and the imbalance between research needs and available resources, the task of establishing priorities and reasonable time-frame expectations seems overwhelming. Actually, the process is not as difficult as it appears at first glance. Within each discipline, the bottlenecks at the subdiscipline level can be identified and research priorities established. For instance, we know that taxonomy represents a major bottleneck in developing IPM. Discussion of Dr. Armbrust's presentation brought out that the taxonomic relationship between the alfalfa weevil and the Egyptian alfalfa weevil is obscure and that biological control efforts have been frustrated because of this fact. In other cases it is not known whether the pest population with which we are dealing is local or represented by migrating individuals.

In biological control our efforts are reduced to a "cut and try" approach unless we understand the phenological cues that regulate diapausing and active stages of host and parasite. By careful and systematic study, the research needed to round out these pieces of the IPM "jigsaw puzzle" can be identified and priorities established. This approach will provide progress on an ad hoc basis, and without a coordinated master plan these contributions will, in time, add to our total resources in support of comprehensive IPM programs. It is largely in this way that progress has been made in the past. This approach is essential but it falls short of testing and advancing the IPM concept. Over and above this need is the larger need for large scale, interdisciplinary effort directed to major crops of regional interest. Such tests will involve high risk efforts in putting the concept to the test and in covering the continuum from research, to demonstration, to practice.

While our attention is directed primarily to research needs of a biological nature, we should not overlook the need for research in the social and economic components. We need to know far more regarding the pattern of pesticide use by growers and factors influencing their decision making. The reports by Beal and his associates (14) and by von Rümker et al. (15) suggest that both farmer and dealer are more dependent on information from the chemical industry than from the public sector.

3. A critical mass of interdisciplinary scientific resources is required to advance IPM. None of the scientific disciplines is sufficiently advanced to provide all that it should in developing IPM programs, so disciplinary work must continue. But, despite this fact, coordinated research must be done by all of the disciplines to develop complete programs. This is an acute need, and it has been only very recently that our concepts were sufficiently developed to identify this need clearly. We have had only one major project that attempted to assemble the resources for IPM of insects on major crops. Fortunately we have had a report on this landmark project from Dr. Carl Huffaker, whose pioneering efforts in organizing the project led to its designation as the "Huffaker Project."

It is now time to extend the IPM concept to other disciplines and assemble the teams qualified to do the work. A successor to the Huffaker Project with the added interdisciplinary dimension has recently been proposed and is designated the Adkisson proposal[1]. This proposal would involve interdisciplinary teams at 17 major land-grant universities for research on all classes of pests of 6 major crops. The proposal is for 5 years at a cost of approximately $5 million per year in addition to the state funds already supporting personnel and facilities that can be allocated to the project. Projects of this scope encounter difficulty in organization, funding, and administration.

Interdisciplinary Cooperation

We have for a hundred years attempted to develop strong disciplines as foundation stones of the university, and the departmental unit as we know it today is the result of this effort. The rewards system works through this unit in matters

[1]*A National Program for the Development of Comprehensive, Unified, Economically and Environmentally Sound Systems of Integrated Pest Management for Major Crops. A Program Project Proposal. Compiled by Dr. Perry Adkisson, Dept. of Entomology, Texas A&M University, College Station, Texas 77843— February 1977.*

of peer recognition, promotion, and salary. We have not yet
found a substitute for the departmental unit, although innovations have been attempted——institutes, centers, task
forces, etc. Another deterrent is the fact that IPM has been
dominated by entomology. The problems of insecticides forced
IPM upon entomologists, and some of the other essential disciplines feel uncomfortable in joining the team with some conditions already established by one discipline. It has been
even more difficult to build economists in from the start,
despite the fact that economic considerations will account for
whether a program is acceptable.

When we consider the several disciplines required and the
philosophical and administrative barriers to interdisciplinary
work, it becomes clear that there are a limited number of institutions that can assemble the critical mass required to
advance IPM programs of this type. The hard decision has to
be made whether we concentrate resources in these teams or
spread resources more evenly but without the essential critical mass.

4. Funding levels for IPM are inadequate and administrative constraints on existing funds limit their effectiveness.
The levels of funding required for pathfinding projects such
as the Adkisson proposal exceed those available from a single
agency. These budgets can be met only by the combined support
of several agencies. A precedent already exists for a combined effort in the Huffaker project, which was funded by
NSF, EPA, and USDA.

But the combining of funds runs counter to traditional
administrative arrangements. NSF is understandably concerned
with basic research, and would therefore look to another
agency for support of applied phases of IPM. USDA recognizes
the need for both basic and applied but finds it difficult to
buy in through its established agencies, Agricultural Research
Service and Cooperative State Research Service. ARS, despite
its reorganization, which should help it meet regional interdisciplinary needs, does not have the range of disciplinary
expertise to assemble the teams.

CSRS is comfortable with the kind of research required,
but is favorable to the system whereby formula funding insures
that the available funds are divided among the state experiment stations.

The Federal Extension Service is expected to play a key
role in IPM, and encounters difficulties in identifying its
role with the research components and in harmonizing its efforts with EPA's aspirations to implement programs.

The outcome of these different factors is that the one
comprehensive project designed to test the concept of IPM

applied to entomology is scheduled to terminate in February, 1978. It is generally conceded that the project has achieved a high order of success. Its logical successor with an added interdisciplinary dimension is apparently stalled, not because it lacks scientific merit but because its funding and administration cannot be neatly fitted to the existing agency structures and administrative procedures. If the existing structures lack the flexibility to meet the need, a new structure might be in order.

EXTENSION NEEDS

Role in Education

The Cooperative Extension Service (CES) is the off-campus educational arm of the land-grant university system. Its organization, financing, and program planning is shared jointly by the USDA, land-grant universities, and community leaders elected or appointed to offices in the county extension organization. CES is a unique institution in its linkage between federal, state, and local agencies. With organizations in almost every county of the United States, a professional staff of about 17,000 individuals, and a tradition of effective service to the agricultural industry of the nation, it represents a major resource in dealing with matters of agricultural production, land use, and public policy.

In recent years there has been an effort to extend the CES delivery system to groups and individuals beyond its traditional agricultural clientele. In principle, CES has an obligation to all segments of the public that have need of its knowledge resources, including environmentalists and others who have no direct economic investment in agriculture.

This brief comment on the role of CES and its modus operandi serves to identify the numerous interfaces and sensitive relationships that exist as CES seeks to function as a nonpartisan educational force in dealing with such complex issues as pesticide use and regulation.

Role in Pest Control

CES plays a key role in implementing the educational programs relating to all phases of pest control. Activities included are—
1. training for certification of users of pesticides as required by FEPCA;
2. providing pesticide recommendations as a guide to

suppliers and users;
3. serving as a source of information to community
   leaders, elected officials, industry, regulatory
   agencies, consultants, producers, and the general
   public;
4. implementing pest management programs.

Role in IPM

CES has several roles in the broad area of IPM. These
include—
1. selling the concept of IPM;
2. providing training for growers, industry representatives, and consultants;
3. implementing programs.
It is always difficult to change existing practices.
The education leadership of CES is needed to establish the
concept of IPM as the guide to the future. The implementation of IPM is largely an educational process, and the land-
grant university with its interacting programs of teaching,
research, and extension is uniquely qualified to provide the
needed leadership.

On the other hand, it must be recognized that CES guards
very jealously its record of credibility with its clientele
and is reluctant to provide programs that it feels involve
undue risk to the producer.

Prerequisites for Success

Recognizing that early failures in IPM programs will set
back the general acceptance of both concept and practice, CES
must proceed cautiously. The prerequisites for success are—
1. adequate knowledge base;
2. integration of components;
3. a delivery system;
4. personnel.
We have already discussed in the section on research the
inadequacy of the knowledge base. But in the case of all the
major crops there is enough knowledge to justify a start. The
caution should not apply to the question of whether or not to
start implementing some features of an IPM program. The
caution must be applied in making certain that the program is
not oversold and that unrealistic expectations are not developed.

The delivery system involves many logistical considerations such as monitoring personnel, turn-around time, computer

resources, and cost. Dr. Haynes has provided us with an excellent overview of the needs for modeling and systems analysis and the conversion of these findings into a program useful to the producer as an aid to his decision making (see p.
181). Dr. Croft has shared with us some encouraging successes
in the application of these methods to the program of IPM on
apple in Michigan (see p. 101).

Special attention must be given to the personnel requirement for the implementation of IPM through CES. The concern
here is that adequate personnel be available with training
equal to the needs. Retread training for established extension personnel is badly needed. Concepts have been changing
at a rapid rate and the personnel system has not provided well
for updating personnel. The training needs could perhaps be
best met on a regional basis through training centers.

Pilot Programs in Pest Management

In 1972 the USDA initiated pilot programs in pest management. The rationale underlying these projects was——
1. the application of existing knowledge could result in
   reduction in use of pesticides with savings to
   growers;
2. that this could be demonstrated to growers;
3. once demonstrated, growers would be willing to share
   in the cost of these programs.
The initial effort in 1972 involved two projects, one in
Arizona on cotton insects, and the other in North Carolina on
tobacco insects. In 1973, 3-year pilot projects were initiated on cotton in 14 states. These programs have been expanded to include corn, soybeans, rice, peanuts, tobacco, wheat,
grain sorghum, alfalfa, potatoes, vegetables, citrus, pears,
apples, peaches, and pecans. By 1976, 33 states were participating in 38 projects with funding of $2,885,000. A progress
report on these projects has recently been provided (16). In
FY 1978, all states will be funded at not less than $25,000
of federal funds with a maximum of $115,000 to states with the
greatest pesticide use. While projects must meet USDA guidelines, they can be on commodities and pests selected by the
individual states.

The ultimate objective of the pilot programs has been to
provide education and technical assistance rather than to
provide farmers with individual services. Pilot programs that
were extended beyond the original 3-year period have required
that growers pay direct costs of scouting fields.

In general, these pilot programs have been highly successful, as has been documented by a number of reports (17-19).

It should be pointed out, however, that a number of these
projects are not IPM programs in the broad sense. In the case
of cotton, for instance, insect supression has been the main
objective, with primary emphasis on insecticide applications
"as needed," based on monitoring of pest levels by scouting.
This fact should not detract from the value of these pilot
studies, but indicates the need for expansion to include more
components of pest control.

What the Pilot Programs Have Demonstrated

In a relatively short period of time, pilot programs have
been implemented and analyzed to provide guidelines for the
future. The conclusions that can be drawn from these studies
are:
1. IPM programs can result in savings to growers through
   reduced use of pesticides.
2. Participation of growers in IPM programs brings about
   a change in outlook regarding the use of pesticides.
3. Growers are willing to bear some of the cost of IPM
   programs once their effectiveness has been demonstrated.
4. IPM programs should be expanded to include:
   a. increased acreage of crops already under IPM;
   b. more components of IPM (to go beyond scouting)
      and pests other than insects;
   c. new crops to further demonstrate the effective-
      ness of IPM.
These expansions will focus on several unresolved issues
including: (1) how the projects will be funded, and (2) what
delivery system will be used.

It is unlikely that funds from the public sector will be
adequate for marked expansion of IPM along the lines of the
pilot studies. It seems reasonable to assume that public
funds will be provided on a temporary basis with the delivery
of pest management technology being placed on a self-support-
ing basis over time.

Funds are inadequate to increase extension personnel to
meet the needs of expanded IPM programs. The transition of
technology delivery from the public to the private sector involves many uncertainties. CES, having demonstrated the effectiveness of pilot programs, is now intensifying its efforts
to transfer the delivery system to grower cooperatives and
pest management consultants. The recent report by Good et al.
(20) provides some guidelines for establishing grower-owned
organizations. A report on the future outlook for IPM as
envisioned by ES-USDA has been provided (21).

Although CES suffered some loss of credibility for its role in overdoing chemical control, it should be remembered that the real deficiency was in the concepts and research base on which recommendations were based. CES is always vulnerable if the research base on which its programs are based is inadequate. In general, CES's record of educational leadership is good and can be improved. In any event, the educational role in advancing IPM should remain with CES. Whatever its shortcomings at this point, it would be better to concentrate on overcoming these than to diffuse responsibility between two or more federal agencies and accept the confusion and territorial disputes that would probably occur.

The era of chemical control disclosed some basic weaknesses that threatened to undermine effectiveness and create environmental problems. The logical successor to chemical control is IPM, which seeks to reconcile pest control with ecological principles. IPM is a concept, a state of mind. Implementation is in its early stages and despite the encouraging reports presented here much needs to be done to strengthen the teaching, research, and extension foundations of IPM. Although the costs in time and resources are high, they are appealing considering the alternatives. The challenge clearly requires reorientation, interdisciplinary effort, new research strategy, additional funding, and new administrative arrangements. IPM is an idea whose time has come.

REFERENCES

1. National Academy of Sciences. 1977. Analytical Studies of U.S. Environmental Protection Agency. Vol. 1, Perspectives on Technical Information for Environmental Protection. NAS, National Research Council, Washington, D.C.
2. National Academy of Sciences. 1972. Report of the Committee on Research Advisory to the U.S. Department of Agriculture. National Research Council. NTIS, Springfield, Virginia.
3. Perkins, J.A. 1966. The University in Transition. Princeton University Press, Princeton, New Jersey.
4. Matooka, P.S. (ed.) 1975. Proceedings: Pest Management Seminar for Agricultural Administrators. East/West Center, Honolulu, Hawaii.
5. Glass, E.H. (ed.) 1975. Integrated Pest Management: Rationale, Potential, Needs and Implementation. Ent. Soc. Am., Spec. Publ. 75-2, College Park, Maryland.
6. Metcalf, R.L. and W.H. Luckmann. 1975. Introduction to Insect Pest Management. John Wiley and Sons, New York.
7. Anonymous. 1977. A professional program leading to the award of the degree of Master of Pest Management. Circ. 7,

Pestology Center, Dept. Biol. Sci., Simon Frasier University, Burnaby, British Columbia.

8. Matooka, P.S. (ed.) 1974. Proceedings Workshop on the Professional Doctorate in Pest Management. East/West Center, Honolulu, Hawaii.

9. Peterson, G.D. 1976. Pest Management: A New Concept Needs a New Educational Approach. East/West Center, Honolulu, Hawaii.

10. Browning, C.B. 1972. Systems of pest management and plant protection. Resident Instruction Committee on Organization and Policy (RICOP), University of Florida, Gainesville, Florida.

11. Obrien, S.R. and P.S. Matooka. (eds.) 1972. Proceedings: Workshop on Pest Management, Curriculum Development and Training Needs. East/West Center, Honolulu, Hawaii.

12. Ventre, A.M. 1975. Current Pest Management Educational Programs and Related Job Opportunities. Office of Pesticide Programs, Environmental Protection Agency, Washington, D.C.

13. National Academy of Sciences. 1975. Pest Control: An Assessment of Present and Alternative Technologies. Vol. I. Contemporary Pest Control Practices and Prospects: the Report of the Executive Committee. Washington, D.C.

14. Beal, G.M., J.M. Bohlen, W.A. Fleishman. 1966. Behavior studies related to pesticides, agricultural chemicals and Iowa farmers. Iowa State Univ. Spec. Rep. No. 49.

15. von Rümker, R. et al. 1974. Farmers' Pesticide Use Decisions and Attitudes on Alternate Crop Protection Methods. Office of Pesticide Programs, Environmental Protection Agency, Washington, D.C.

16. Good, J.M. 1977. Progress Report. Pest Management Plot Projects. Ext. Serv. USDA, Washington, D.C.

17. von Rümker, R. et al. 1975. Evaluation of Pest Management Programs for Cotton, Peanuts and Tobacco in the United States. Office of Pesticide Programs, Environmental Protection Agency (EPA-540/9-75-031), Washington, D.C.

18. Good, J.M. and N.P. Bonell. 1976. Cotton Insect Pest Management Plot Projects 1972-74. Ext. Serv., USDA, Agriculture and Natural Resources-5-64 (4-76).

19. Blair, B.D. 1976. Extension pest management. U.S. Federal Ext. Serv., ESC-579, USDA.

20. Good, J.M., R.E. Hepp, P.O. Mohn, D.L. Vogelsang. 1977. Establishing and Operating Grower-Owned Organizations for Integrated Pest Management. Ext. Serv., USDA, Program Aid-1180.

21. Good, J.M. 1977. Integrated pest management—a look to the future. U.S. Federal Ext. Serv., ESC-853, USDA.

INDEX